JN119583

Study with your Friends!

How do we learn mathematics?

Based on the problem you find in your daily life or what you have learned, let's come up with a purpose.

1

The first problem of the lesson is written. On the left side, what you are going to learn from now on through the problem is written.

When you see the problem and think that you "want to think", "want to represent", "want to know", and "want to explore", that will be your "purpose" of your learning. You can find the purpose not only at the beginning of the lesson but in various timings and settings.

You can check your understanding and try more using what you have learned.

①

Let's try this problem first.

The starting point

Find the ?

Can we share the cookies equally?

We have 12 cookies. Let's divide them.
Among how many people?
How many each?
There are 12 cookies. Let's share
There are 12 cookies. Let's share

What you have learned today

3 Division

Let's think about how to divide things equally.

① Calculation to find out the number for each child

Finding out the number

1 If you divide 12 cookies equally among 4 children, how many cookies does each child get?

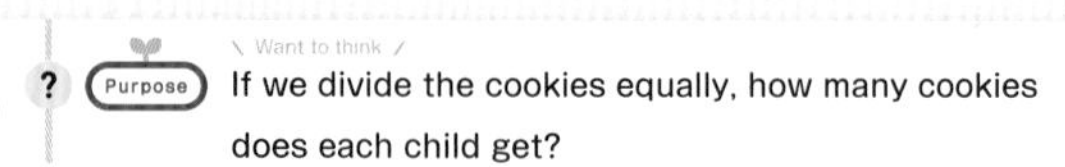

? Purpose Want to think: If we divide the cookies equally, how many cookies does each child get?

Purpose Can you apply the rules you have learned so far?

1 Let's write a math sentence for each situation below and find out the number of blocks given to each child.

① Divide 6 blocks equally among 3 children.

☐ ÷ ☐ = ☐

Total number / Number of children / Number of blocks for each child

1 Let's calculate the following in vertical form.

① 153 + 425 ② 261 + 637 ③ 437 + 302 ④ 502 + 207

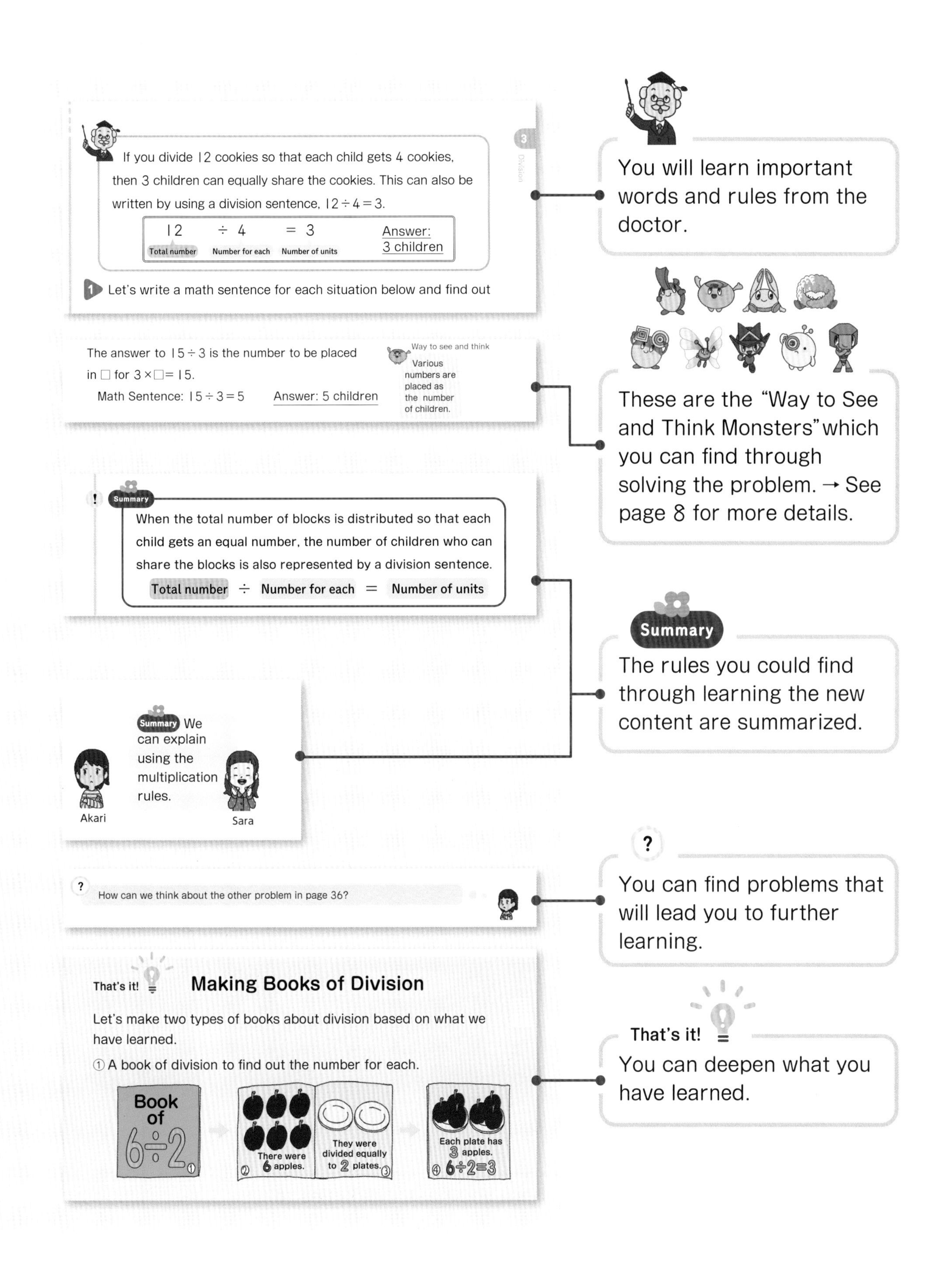

If you divide 12 cookies so that each child gets 4 cookies, then 3 children can equally share the cookies. This can also be written by using a division sentence, 12 ÷ 4 = 3.
12 ÷ 4 = 3
Total number
Number for each
Number of units
Answer: 3 children
1 Let's write a math sentence for each situation below and find out
You will learn important words and rules from the doctor.
The answer to 15 ÷ 3 is the number to be placed in □ for 3 × □ = 15.
Math Sentence: 15 ÷ 3 = 5
Answer: 5 children
Way to see and think
Various numbers are placed as the number of children.
These are the "Way to See and Think Monsters" which you can find through solving the problem. → See page 8 for more details.
Summary
When the total number of blocks is distributed so that each child gets an equal number, the number of children who can share the blocks is also represented by a division sentence.
Total number ÷ Number for each = Number of units
Summary
The rules you could find through learning the new content are summarized.
Summary We can explain using the multiplication rules.
Akari
Sara
?
How can we think about the other problem in page 36?
?
You can find problems that will lead you to further learning.
That's it!
Making Books of Division
Let's make two types of books about division based on what we have learned.
① A book of division to find out the number for each.
Book of 6÷2
There were 6 apples.
They were divided equally to 2 plates.
Each plate has 3 apples.
6÷2=3
That's it!
You can deepen what you have learned.

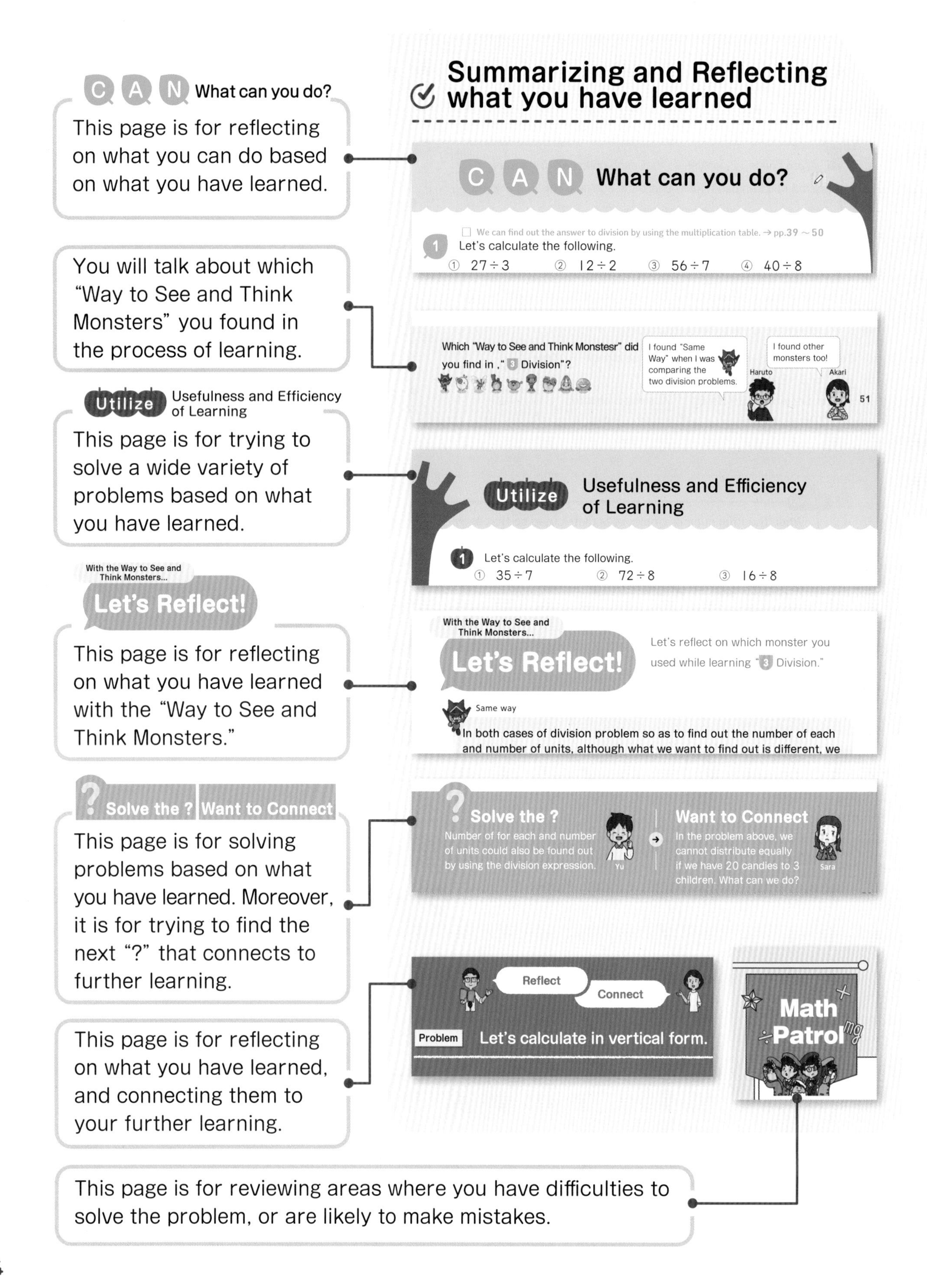

Summarizing and Reflecting what you have learned
CAN What can you do?
This page is for reflecting on what you can do based on what you have learned.
CAN What can you do?
We can find out the answer to division by using the multiplication table. → pp.39 ~ 50
1 Let's calculate the following.
① 27÷3 ② 12÷2 ③ 56÷7 ④ 40÷8
You will talk about which "Way to See and Think Monsters" you found in the process of learning.
Which "Way to See and Think Monstesr" did you find in ," 3 Division"?
I found "Same Way" when I was comparing the two division problems.
Haruto
I found other monsters too!
Akari
51
Utilize Usefulness and Efficiency of Learning
This page is for trying to solve a wide variety of problems based on what you have learned.
Utilize Usefulness and Efficiency of Learning
1 Let's calculate the following.
① 35÷7 ② 72÷8 ③ 16÷8
With the Way to See and Think Monsters...
Let's Reflect!
This page is for reflecting on what you have learned with the "Way to See and Think Monsters."
With the Way to See and Think Monsters...
Let's Reflect!
Let's reflect on which monster you used while learning "3 Division."
Same way
In both cases of division problem so as to find out the number of each and number of units, although what we want to find out is different, we
Solve the ? Want to Connect
This page is for solving problems based on what you have learned. Moreover, it is for trying to find the next "?" that connects to further learning.
Solve the ?
Number of for each and number of units could also be found out by using the division expression.
Yu
Want to Connect
In the problem above, we cannot distribute equally if we have 20 candies to 3 children. What can we do?
Sara
This page is for reflecting on what you have learned, and connecting them to your further learning.
Reflect
Connect
Problem Let's calculate in vertical form.
Math Patrol
This page is for reviewing areas where you have difficulties to solve the problem, or are likely to make mistakes.

Utilizing Math for SDGs

Watch out for traffic accidents!

Utilizing Math for SDGs

The Sustainable Development Goals (SDGs) are a set of goals that we aim to achieve in order to create a world where we can live a life of safety and security. This page will help you think about what you can do for society and the world through math.

About the QR Code

Some of the pages include the QR code which is shown on the right.

… You can learn how to draw a diagram and how to calculate by watching a movie.

… You can learn by actually moving and operating the contents.

… You can learn by reflecting on what you have learned previously in your previous grades.

… You can utilize it to know the solution to the problems that you couldn't find out the answer, or to try various problems.

… You can deepen your learning by actually looking at the materials including the website.

Dear Teachers and Parents

This textbook has been compiled in the hope that children will enjoy learning through acquiring mathematical knowledge and skills.
The unit pages are carefully written to ensure that students can understand the content they are expected to master at that grade level.
In addition, the "More Math!" section at the end of the book is designed to ensure that each student has mastered the content of the main text, and is intended to be handled selectively according to the actual conditions and interests of each child.
We hope that this textbook will help children develop an interest in mathematics and become more motivated to learn.

The sections marked with this symbol deal with content that is not presented in the Courses of Study for that grade level, thus does not have to be studied uniformly by all children.

QR codes are used to connect to Internet content by launching a QR code-reading application on a smartphone or tablet and reading the code with a camera.
The QR Code can be used to access content on the Internet.

The code can also be used at the address below.
https://r6.gakuto-plus.jp/s3a0l

Note: This book is an English translation of a Japanese mathematics textbook. The only language used in the contents on the Internet is Japanese.

【Infectious Disease Control】
In this textbook, pictures of activities and illustrations of characters do not show children wearing masks, etc., in order to cultivate children's rich spirit of communicating and learning from each other. Please be careful to avoid infectious diseases when conducting classes.

Becoming a Writing Master

The notebook can be used effectively.

- To organize your own thoughts and ideas
- To summarize what you have learned in class
- To reflect on the what you have learned previously

Let's all try to become notebook masters.

Write today's date. → November 16th

Write the problem of the day that you must solve. → Problem

Let's find out the answers of 3 × 10, 3 × 11, and 3 × 12.

Let's write down what you thought while thinking about the solution of the problem as "purpose." → ⟨Purpose⟩ Can we use the rules we have found so far?

Write your ideas or what you found about the problem. → ○ My idea

The answer for the row of 3 in the multiplication table increases by 3.

3 × 9 = ~~26~~ 27) + 3
3 × 10 = 30) + 3
3 × 11 = 33) + 3
3 × 12 = 36

Tips for Writing 1

By aligning × and =, it is easier to see.

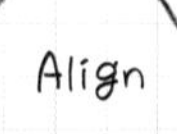

Did I have a similar case before?

Tips for Writing ❶

When you made a mistake, don't erase it so that it will be easier to understand when you look back at your notebook later.

Tips for Writing ❷

By finding the "Way to See and Think Monsters," it will connect you to what you have learned previously.

Tips for Writing ❸

By writing down what you would like to try more, it will lead you to further learning.

○ Yu's idea

He was trying to find out the answer by decomposing 3×10 into 3×5 and 3×5.

3×5 3×5

$3 \times 5 = 15$、 $15 + 15 = 30$

So

$3 \times 10 = 30$

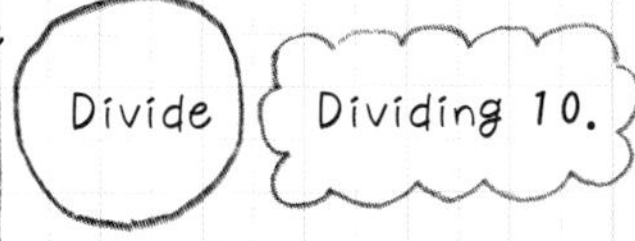

Did I have a similar case before?
Divide
Dividing 10.

Tips for Writing ❷

〈Summary〉

Even if the multiplier is greater than 10, I can find the answer using the rules I have discovered so far.

〈Reflection〉

Evev if we cannot use the multiplication table. I found out that we can do multiplication by using the rules.

〈What I want to do next〉

Can I do multiplication with larger numbers?

Tips for Writing ❸

Write the classmate's ideas you consider good.

Draw a diagram to understand the situation easily.

Summarize what you have learned today.

Reflect on your class, and write down the following;
- What you learned
- What you found out
- What you can do now
- What you don't know yet

While learning mathematics...

Based on what I have learned previously...

Why does this happen?

There seems to be a rule.

You may be in situations like above. In such case, let's try to find the "Way to See and Think Monsters" on page 9. The monsters found there will help you solve the mathematics problems. By learning together with your friend and by finding more "Way to See and Think Monsters," you can enjoy learning and deepening mathematics.

What can we do at these situations?

I think I can use 2 different monsters at the same time...

→ You may find 2 or 3 monsters at the same time.

I came up with the way of thinking which I can't find on page 9.

→ There may be other monsters than the monsters on page 9. Let's find some new monsters by yourselves.

Now let's open to page 9 and reflect on the monsters you found in the 2nd grade. They surely will help your mathematics learning in the 3rd grade!

Representing ways of thinking in mathematics

Way to See and Think Monsters

Unit

If you set the unit...

Once you have decided one unit, you can represent how many using the unit.

Summarize

If you try to summarize...

It makes it easier to understand if you summarize the numbers or summarize in a table or a graph.

Other Way

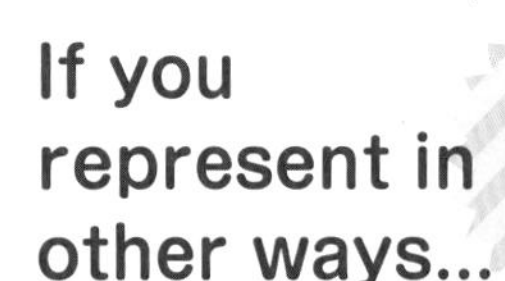

If you represent in other ways...

If you represent in other something depending on your purpose it is easier to understand.

Align

If you try to align...

You can compare if you align the number place and align the unit.

Change

If you try to change the number or the figure...

If you try to change the problem a little, you can understand the problem better or find a new problem.

Divide

If you try to divide...

Decomposing numbers by place value and dividing figures makes it easier to think about problems.

Why

You wonder why?

Why does this happen? If you communicate the reasons in order, it will be easier to understand for others.

Rule

Is there a rule?

By examining, you can find rules and think using rules.

Same Way

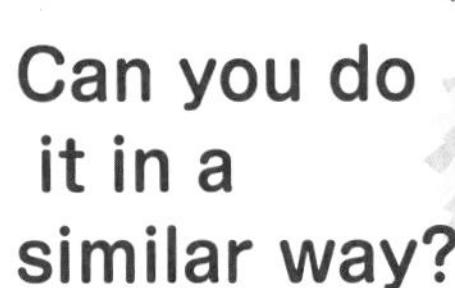

Can you do it in a similar way?

If you find something the same or similar to what you have learned, you can understand.

Ways to think learned in the 2nd grade

Numbers and Calculations

Align / Same Way

Calculation with 1-digit numbers you have learned in the 1st grade can be used by aligning the digits of the numbers.

$$\begin{array}{r} 24 \\ +\ 13 \\ \hline 37 \end{array}$$

2 + 1 = 3　　4 + 3 = 7

Unit / Same way

By setting a unit, we can represent in the same way as numbers larger than 100.

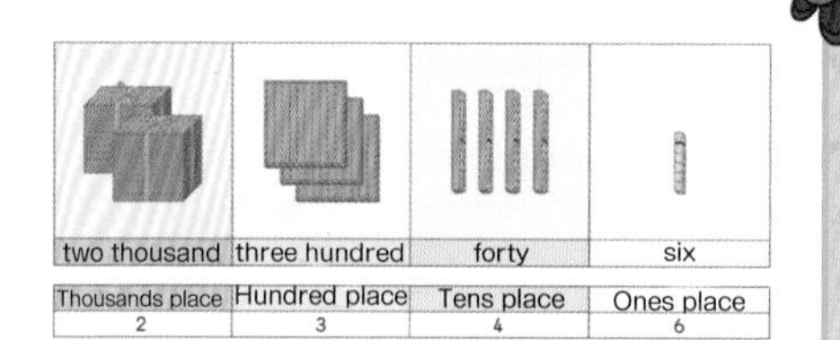

two thousand	three hundred	forty	six
Thousands place	Hundred place	Tens place	Ones place
2	3	4	6

Unit

We can find out the total number by setting one unit.

4 × 3 = 12

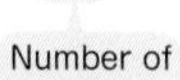

Number of children in each cart / Number of carts / Total number

Rule

We can find out the rules by looking at the multiplication table.

Summary

◯ In multiplication, if the multiplier increases by 1, the answer increases by the multiplicand.
$4 \times 4 = 4 \times 3 + 4$

◯ In multiplication, the answer is the same even if the order of the multiplicand and the multiplier is switched.
$3 \times 5 = 5 \times 3$

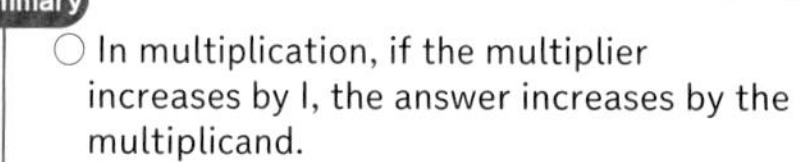

Unit

We can make another unit by dividing the original size equally.

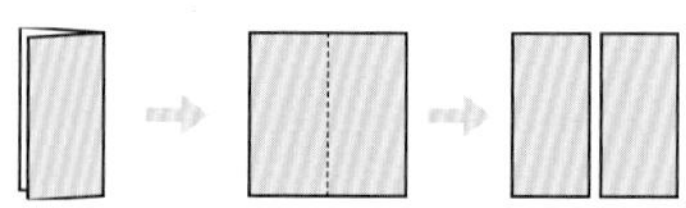

Change / Other Way

We can use addition and subtraction in various situations.

We can represent the problem using diagrams.

in total: ☐ ribbons

Blue: 17ribbons　Red:24ribbons

Shapes

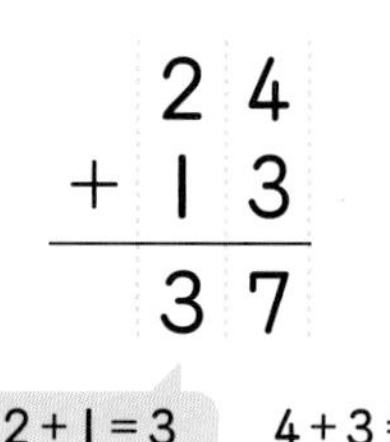

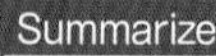

Summarize

Shapes can be classified by the number of sides or vertices.

Rule

We found properties of the shapes of boxes such as the number of sides, faces, vertices and as the shape of faces.

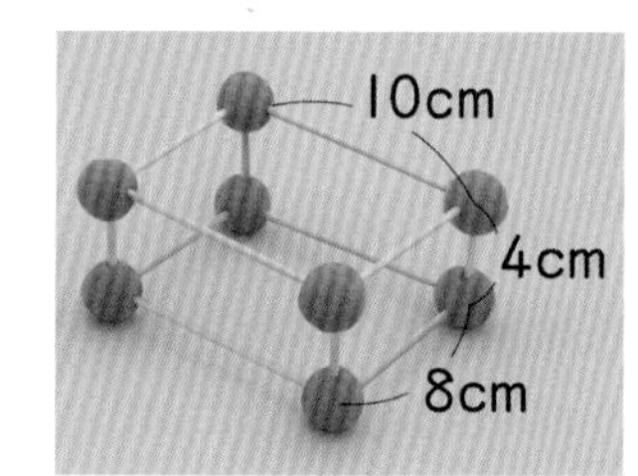

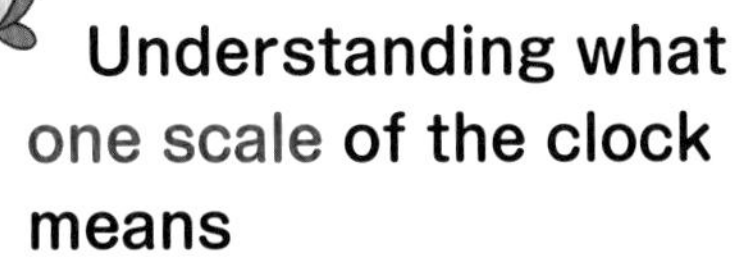

Measurements

Unit

Understanding what one scale of the clock means
- 1 hour = 60 minutes
- 1 day = 24 hours

Other Way

Represent the clock in a diagram.

Unit

Setting 1mm or 1L as one unit.

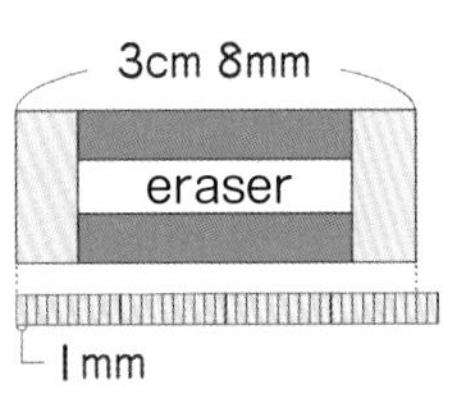

Other Way

"1L = 10dL", "1cm = 10mm" have a similar relationship.

Align

We can calculate length and amount by aligning the digits of the numbers according to their places.

	cm	mm
	4	2
+	3	6
	7	8

Data

Why

Be interested in what you want to investigate.

Other Way

Representing in tables or garphs.

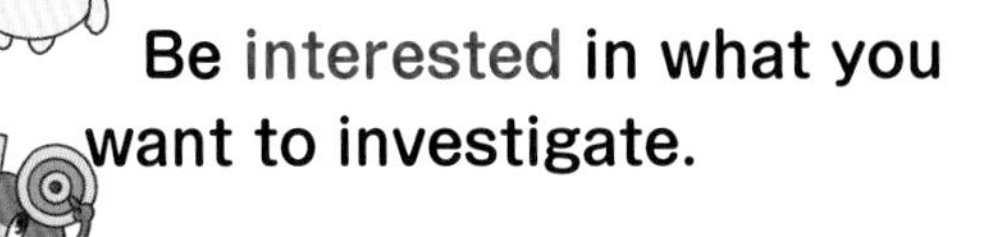

The vegetable I want to grow

Vegetables	Cherry Tomatoe	Cucumber	Soybean	Eggplant	Pepper	Bitter Gourd
Number of Children	7	6	5	3	2	2

The vegetable I want to grow

Cherry Tomato	Cucumber	Soybean	Eggplant	Pepper	Bitter Grourd
○					
○	○				
○	○	○			
○	○	○			
○	○	○	○		
○	○	○	○	○	○
○	○	○	○	○	○

Summarizing depending on your focus.

Summarize

Forgot the multiplication table?

We learned the multiplication table in 2nd grade, but I'm not sure if I can say the row of 7...
1
You're not good at the row of 7?
Yes... Especially I'm not good at 7 × 6 because it's confusing...
2
What can I do if I forget the multiplication table?
You can use the rules of multiplication.
3
Which row are you good at?
I'm okay up until the row of 6...
4
If you can say the row of 6, we can use the diagram of 6 × 7.
5
If you can say the rows of 2 and 5, you can use the diagrams of 2 × 6 and 5 × 6.
6

How can we explain the ways to find out the answer for the multiplication table?

1 Multiplication

Let's find out the rules of multiplication and apply them beyond 9 × 9.

1 The Rules of Multiplication

1 Look at the multiplication table below and find out the answer for 7 × 6 using diagrams or math expressions. 

Multiplier (across) / Multiplicand (down)

	1	2	3	4	5	6	7	8	9
1	1	2	3	4	5	6	7	8	9
2	2	4	6	8	10	12	14	16	18
3	3	6	9	12	15	18	21	24	27
4	4	8	12	16	20	24	28	32	36
5	5	10	15	20	25	30	35	40	45
6	6	12	18	24	30	36	42	48	54
7									
8									
9									

Way to see and think
What rules can you find in the multiplication table?

① Write all the answers in the table above.

The way to find the answer for 7 × 6 is...

Sara

Yu

According to the multiplication table...

② Akari, Yu, and Sara used diagrams and math expressions to think about the way to find the answer for 7×6. Let's explain how they thought.

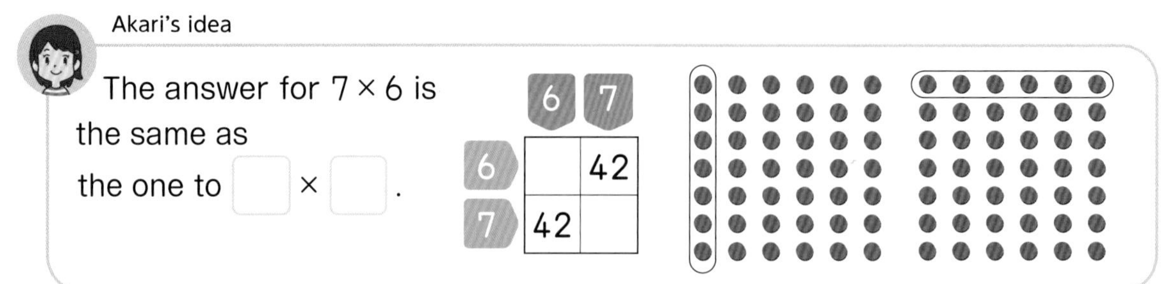

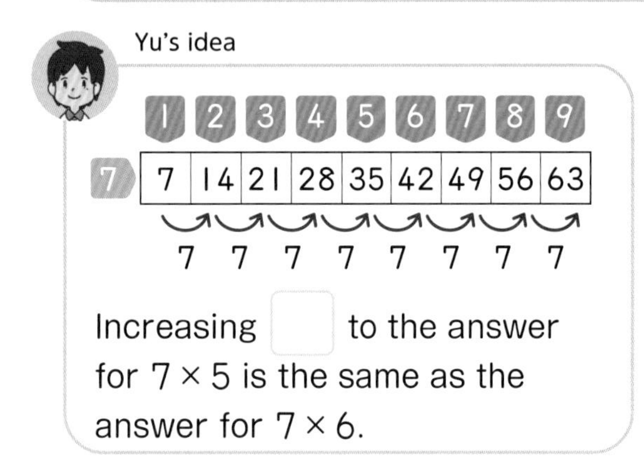

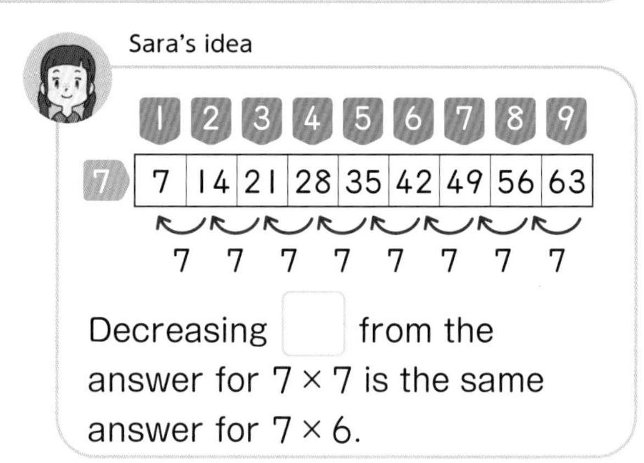

③ When 7×6 is expressed in other math expressions by using the ideas of Akari's, Yu's, and Sara's, each of them can be written as the following. Let's fill in each ☐ with numbers.

Akari $7 \times 6 = 6 \times$ ☐

Yu $7 \times 6 = 7 \times 5 +$ ☐

Sara $7 \times 6 = 7 \times 7 -$ ☐

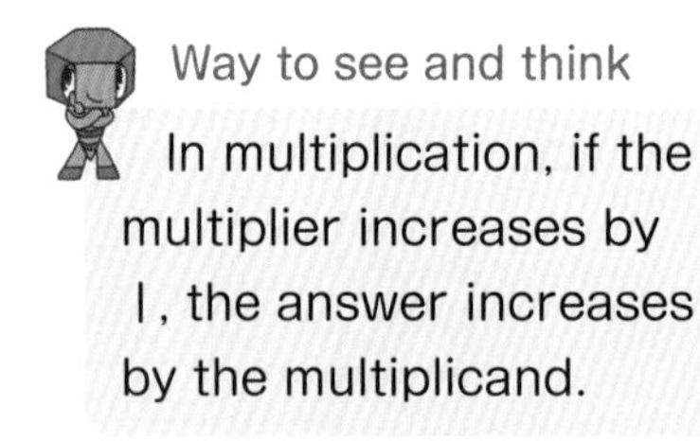

Way to see and think
In multiplication, if the multiplier increases by 1, the answer increases by the multiplicand.

The symbol "=" is called the **equality sign**. It is not only used for writing the answer to any operation, but also for showing that the expressions or the size of numbers on the left side and the right side are equal.

? Are there other ways of thinking which can be found from the multiplication table?

2 In 7 × 6, let's think about what will happen to the answer if you decompose the multiplicand or the multiplier.

① Let's explain the ideas of Haruto's and Akari's.

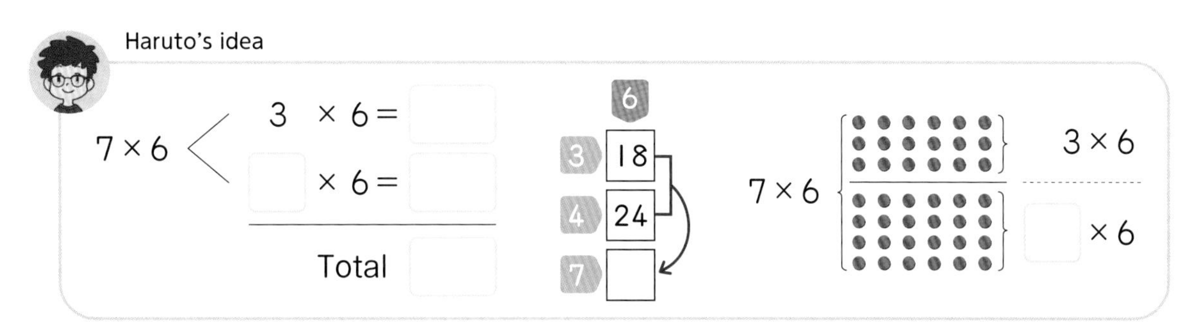

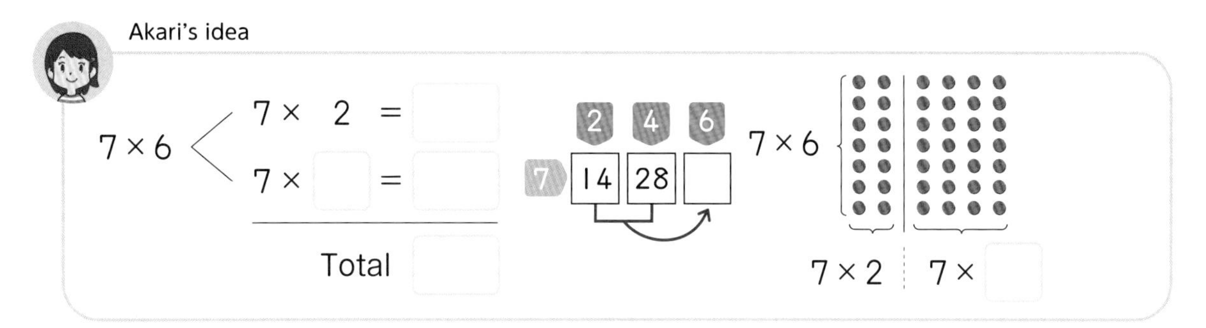

② Yu thought about the calculation of 7×6 as the following. Let's fill in the ☐ with numbers and talk about Yu's idea.

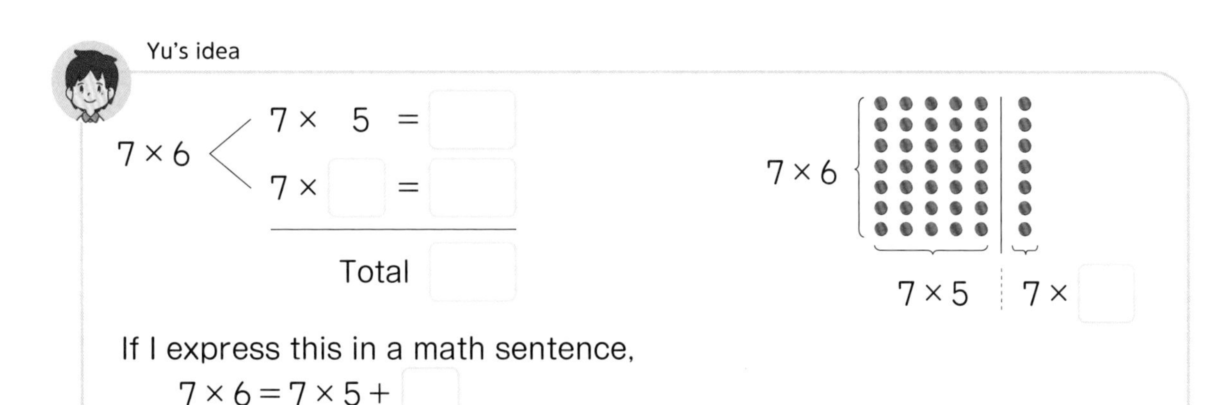

If I express this in a math sentence,

$7 \times 6 = 7 \times 5 + \square$

So, it is the same as the rule of multiplication to calculate by decomposing the multiplier into 1 and some in the rule of multiplication.

③ Let's think how to decompose 7×6 in a different way from ① and ②.

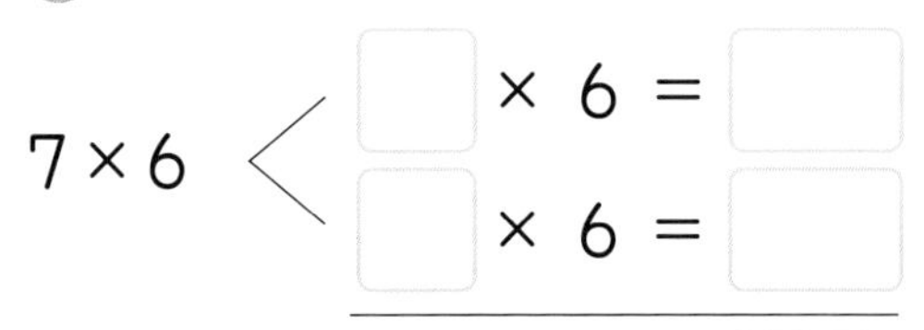

7×6 < $7 \times \square = \square$, $7 \times \square = \square$

Total $\square$

Summary

Based on the rule of multiplication, the answer for 7×6 can be found in various ways.

? Can we apply the same rule of multiplication to multiplications using other numbers?

3

Based on the rules of multiplication which we have found, let's explain the ways of finding out the answer for 8×7.

We found a lot of rules of multiplication.

Sara

Can we use them in other multiplications?

Haruto

\ Want to find out /

Purpose **Is there a multiplication rule which can be applied in any multiplication?**

① Akari explained the multiplication of 8×7 as follows. What multiplication rule did she use?

Akari's idea

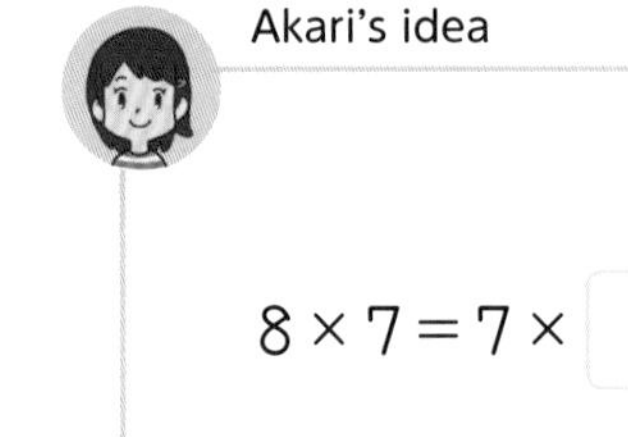

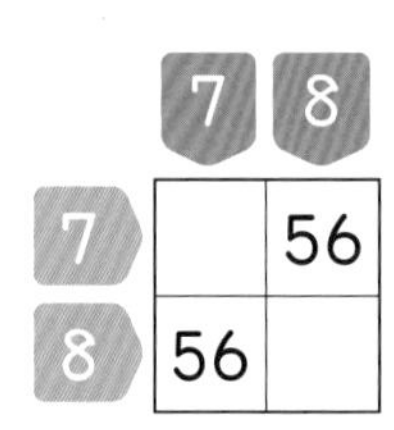

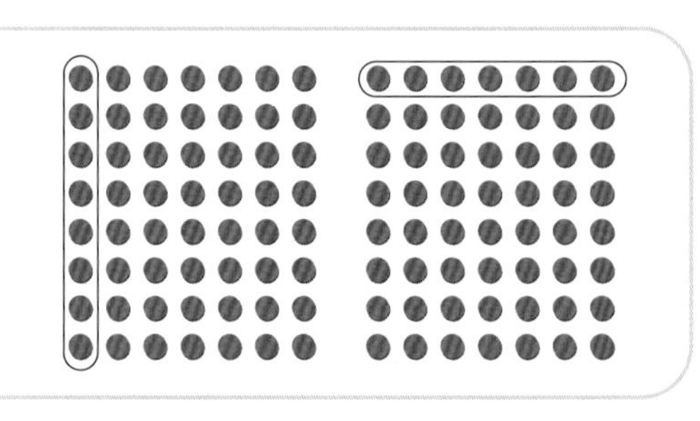

Rules of multiplication →

Summary 【Rule of Commutativity】

In multiplication, the answer is the same even if the order of the multiplicand and the multiplier is switched.

② Yu explained the multiplication of 8 × 7 as follows. What multiplication rule did he use?

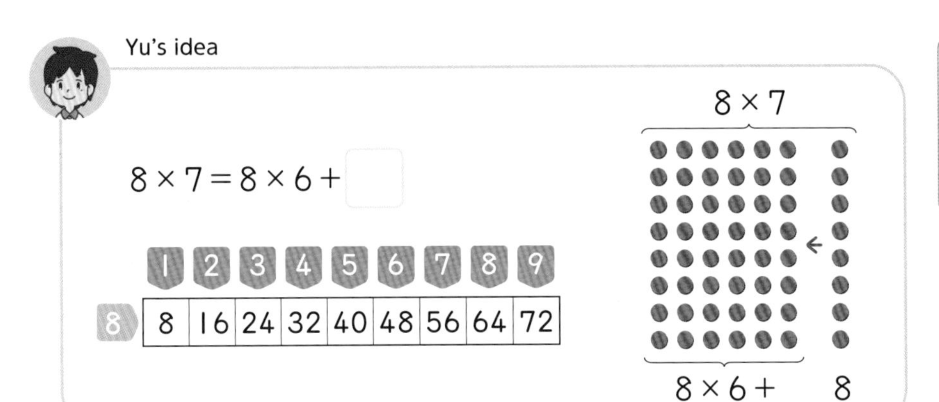

Yu's idea
8 × 7 = 8 × 6 + ☐
1 2 3 4 5 6 7 8 9
8 | 8 16 24 32 40 48 56 64 72
8 × 7
8 × 6 + 8

This is also the same as when we decrease 8 from 8 × 8.

Summary 【Rule of the Multiplier and the Answer】

In multiplication, if the multiplier increases by 1, the answer increases by the multiplicand. Also, if the multiplier decreases by 1, the answer decreases by the multiplicand.

③ Haruto explained the multiplication of 8 × 7 as follows. What multiplication rule did he use?

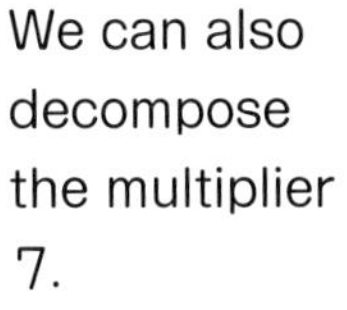

Haruto's idea

Decompose the multiplicand 8 into...

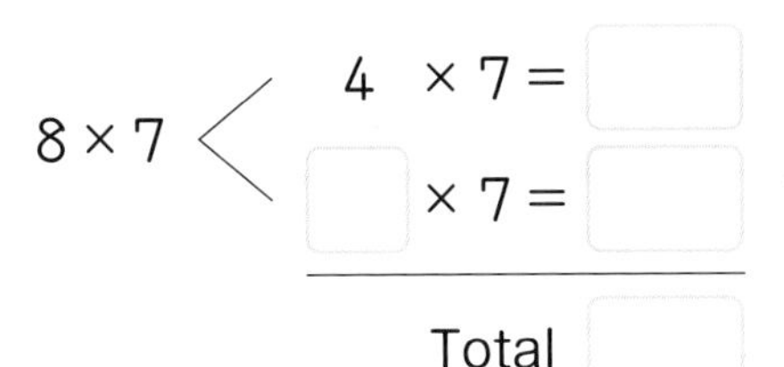

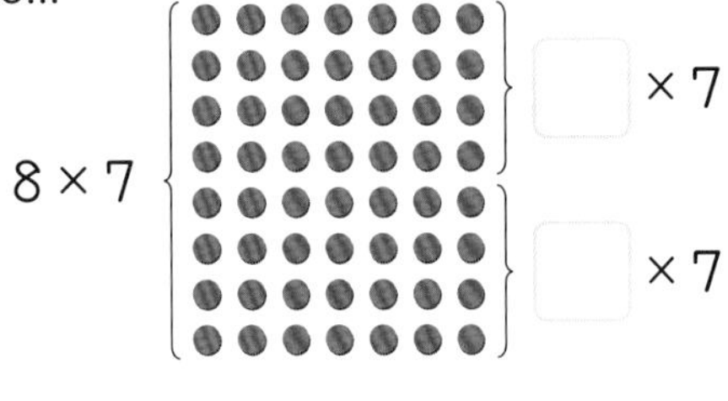

Summary 【Rule of Distribution】

In multiplication, we get the same answer even if we decompose the multiplicand or the **multiplier**.

1 Let's fill in each ☐ with a number.

Way to see and think

Which rule should we apply?

① 4×6 is larger than 4×5 by ☐.

② 5×8 is smaller than 5×9 by ☐.

③ $7 \times 7 = 7 \times \square + 7$

④ $3 \times \square = 3 \times 7 - 3$

⑤ $5 \times 7 = \square \times 5$

⑥ $9 \times \square = 3 \times 9$

2 Let's fill in each ☐ with a number.

① 7×9 < $5 \times 9 = \square$, $\square \times 9 = \square$ — Total ☐

② 4×8

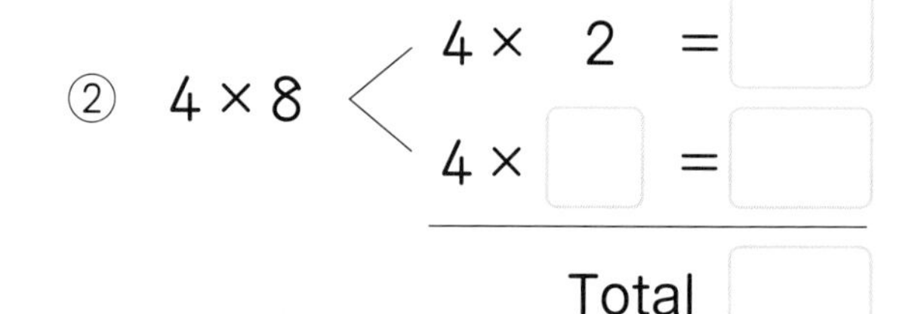

Rule of associativity →

4

Each child gets 2 sticks of 3 dumplings. How many dumplings are needed for 4 children?

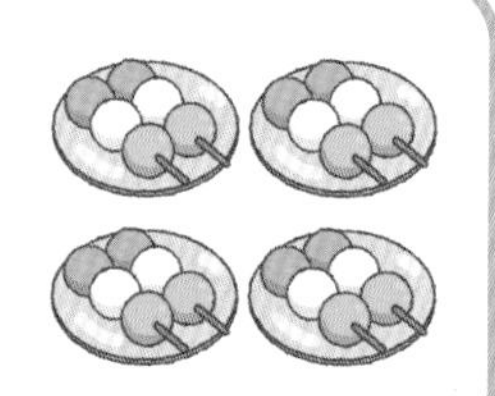

① Sara and Yu found out the answer as shown below. Let's explain how Sara and Yu thought.

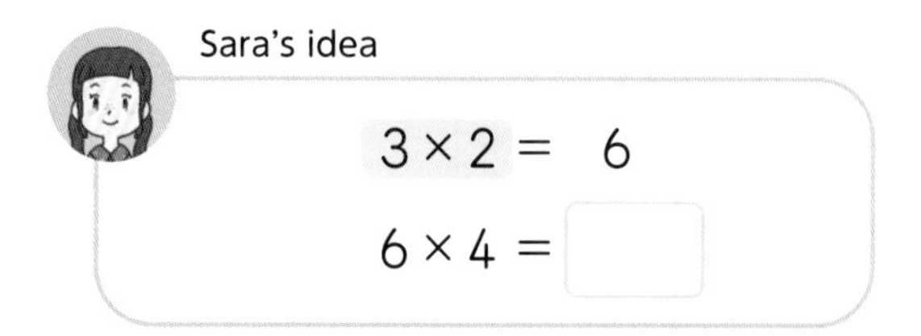

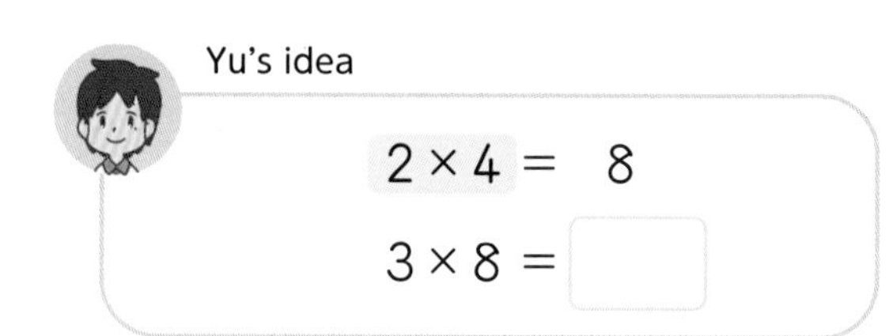

The order of their calculation is different.

What is the way of their thinking?

\ Want to explore /

? **Purpose** **In multiplication, is the answer the same in whatever order we multiply the numbers?**

② Let's express each idea in a math sentence.

Number of dumplings for each child

$(3 \times 2) \times 4 = 24$

Number of sticks

$3 \times (2 \times 4) = 24$

$$(3 \times 2) \times 4 = 3 \times (2 \times 4)$$

Way to see and think

Is it the same way of thinking as adding three numbers?

1 Let's calculate the following and check whether the answer to Ⓐ is same as the answer to Ⓑ.

① Ⓐ $(2 \times 3) \times 3$ Ⓑ $2 \times (3 \times 3)$

② Ⓐ $(2 \times 4) \times 2$ Ⓑ $2 \times (4 \times 2)$

! **Summary** 【Rule of Associativity】

In multiplication, the answer is the same in whatever order we multiply the numbers.

The math expression inside the ( ) should be done first.

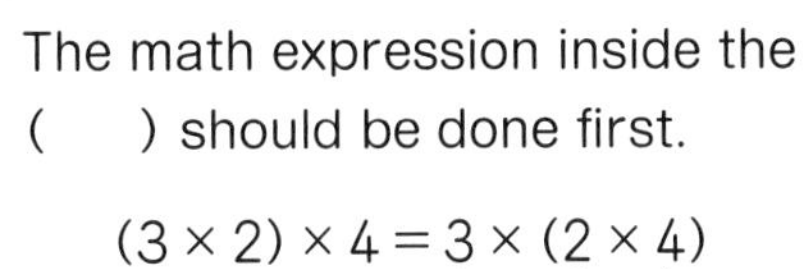

2 Let's calculate the following by changing the order of the numbers.

① $2 \times 2 \times 3$ ② $4 \times 2 \times 4$

3 Sara thought 12×3 as noted on the right. Let's explain how she thought.

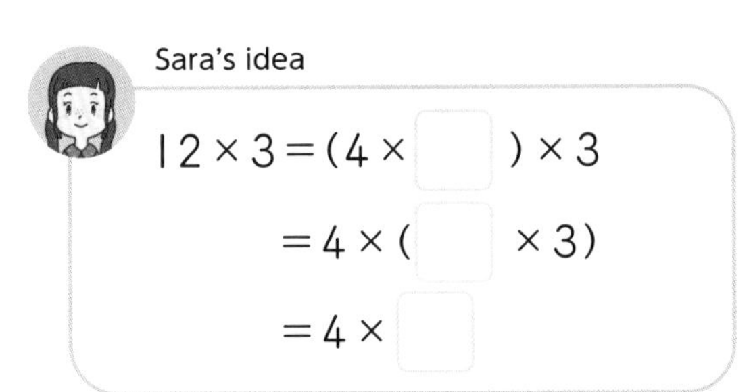

2 Multiplication with 0

Point Scoring Game

You are given 10 marbles to shoot at the target. The points that you can get depends on where the marbles land. The one who gets the highest score wins.

1 The result of Sota's game is shown on the right. Let's see his score.

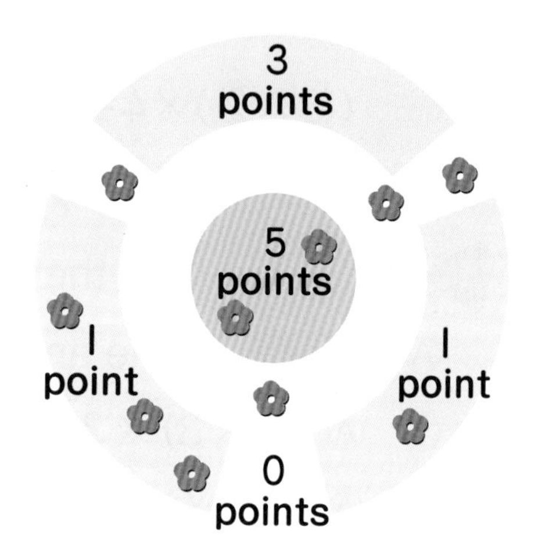

Points	5	3	1	0	Total
Number of marbles					
Score					

① Let's write the math expressions to find out the total points.

	Points		Number of marbles
5-point: 2 marbles …………	5	×	☐
3-point: 0 marbles …………	3	×	☐
1-point: 4 marbles …………	1	×	☐
0-point: 4 marbles …………	0	×	☐

Haruto: 0×4 is $0+0+0+0$, so...

Akari: Can I use any rules of multiplication?

\ Want to explore /

Purpose When the multiplier or the multiplicand is 0 in multiplication, what would the answer be?

② Let's find out the score for 3-point by using the row of 3 in the multiplication table.

$3 \times 4 = 12$
$3 \times 3 = 9$
$3 \times 2 = 6$
$3 \times 1 = 3$
$3 \times 0 = \square$

decreased by $\square$

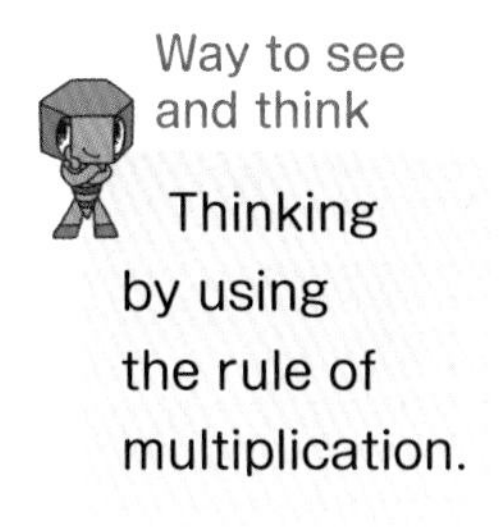

Way to see and think
Thinking by using the rule of multiplication.

③ Let's find out the score for 0-point based on $1 \times 4 = 4$.

$1 \times 4 = 4$
$0 \times 4 = \square$

$\square$ decreased

Way to see and think
Can we apply the rules of multiplication we have learned?

④ Let's complete the table on the left page and find out the total score of Sota.

1 In the point scoring game, what does the math expression 0×0 mean?

Summary

Any number multiplied by 0 equals 0. $3 \times 0 = 0$
Also, 0 multiplied by any number equals 0. $0 \times 4 = 0$
$0 \times 0 = 0$

2 Let's calculate the following.

① 6×0 ② 4×0 ③ 0×7 ④ 0×1

3 In the multiplication table on page 149, let's write the answers in the multiplicands of 0 and in the multipliers of 0.

3 Multiplication with 10

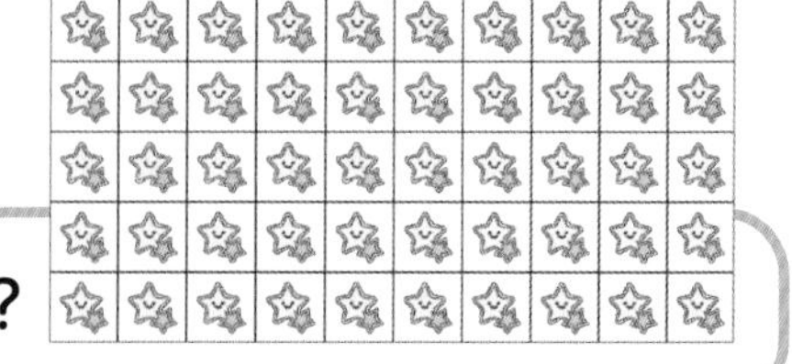

1 How many stickers are there in total?

① Let's write 2 math expressions to find out the number of stickers.

□ × □

□ × □

\ Want to think /

Purpose Can you apply the rules you have learned so far?

Yu

② Let's think about how to find out the answer to 5 × 10 based on the rules of multiplication.

Sara's idea

I used the rule of the multiplier and the answer.

$5 \times 10 = 5 \times 9 + \square = \square$

Yu's idea

I used the rule of distribution.

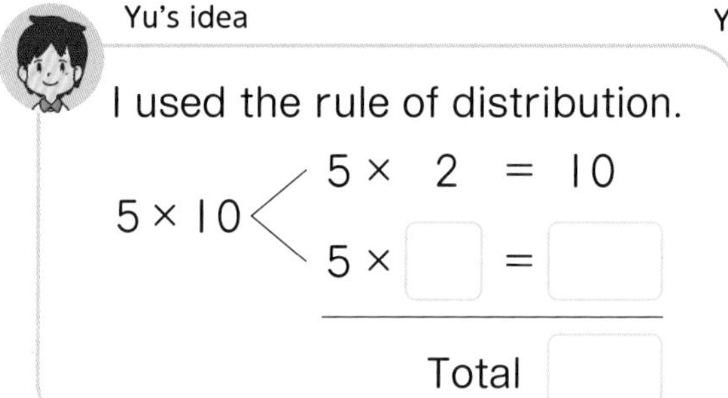

5×10 < $5 \times 2 = 10$, $5 \times \square = \square$

Total □

③ Let's think about how to find out the answer to 10 × 5 based on the rules of multiplication.

Haruto

If we decompose 10 into 7 and 3, I will get 7 × 5 and 3 × 5...

If we use the rule of commutativity, 10 × 5 = 5 × 10···

Akari

Summary We can explain using the multiplication rules.

Sara

1 Let's discuss how to find out the answer to 10 × 10 using Sara's idea in 1 - ②.

Let's write the answer into the table on page 149.

2 Let's calculate the following.

① 6 × 10 ② 8 × 10 ③ 10 × 4 ④ 10 × 9

3 There are 7 children and each one is given 10 stickers. How many stickers are needed in total?

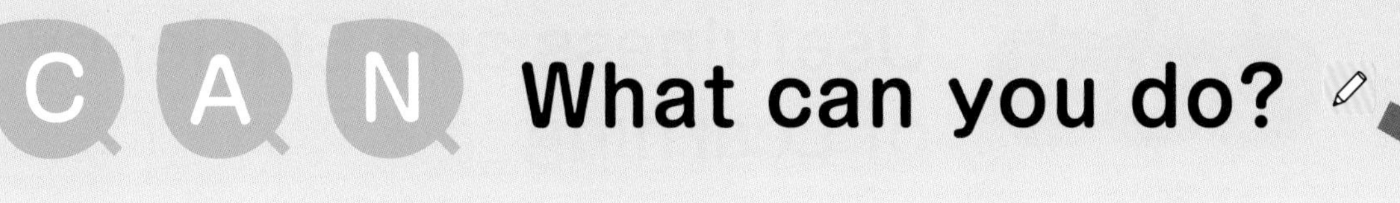

☐ We understand the rule of commutativity. → p.17

1 Let's fill in each ☐ with a number.

① $3 \times 8 = 8 \times$ ☐ ② $4 \times$ ☐ $= 6 \times 4$

☐ We understand the rule of the multiplier and the answer. → p.17

2 Let's fill in each ☐ with a number or an expression.

① In the row of 5, if the multiplier increases by 1, the answer increases by ☐.

② In the row of 9, if the multiplier decreases by 1, the answer decreases by ☐.

③ 3×9 is larger than 3×8 by ☐.

Show in math sentence. $3 \times 9 =$ ☐.

④ 4×3 is smaller than 4×4 by ☐.

Show in math sentence. $4 \times 3 =$ ☐.

☐ We understand the rule of distribution. → p. 18

3 Let's fill in each ☐ with a number.

① 8×7 < $8 \times 3 =$ ☐, $8 \times$ ☐ $=$ ☐; Total ☐

② 9×6 < $5 \times 6 =$ ☐, ☐ $\times 6 =$ ☐; Total ☐

☐ We understand the rule of associativity. → p.19

4 Let's fill in each ☐ with a number.

① $(3 \times 3) \times 2 = 3 \times ($ ☐ $\times 2)$ ② $(7 \times 2) \times 4 = 7 \times (2 \times$ ☐ $)$

Supplementary Problems → p.129

Which "Way to See and Think Monsters" did you find in "1 Multiplication"?

I found "Rule" while working on the multiplication chart.

Akari

I found other monsters too!

Yu

Utilize Usefulness and Efficiency of Learning

1 Parts of the multiplication table are shown below. Let's fill in the numbered spaces with answers.

①

		❸	
❶	30	❷	40
			❹

②

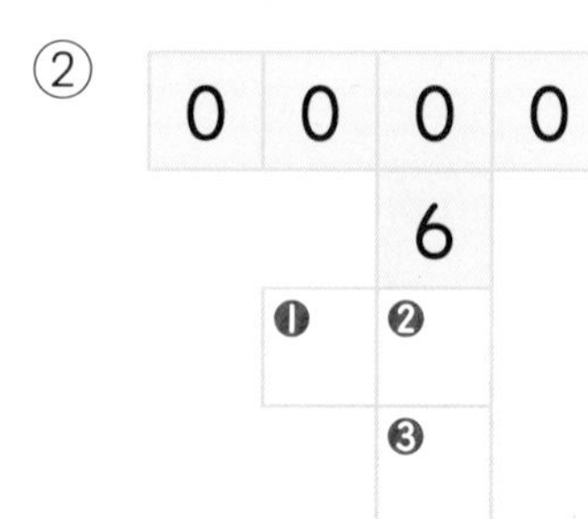

③

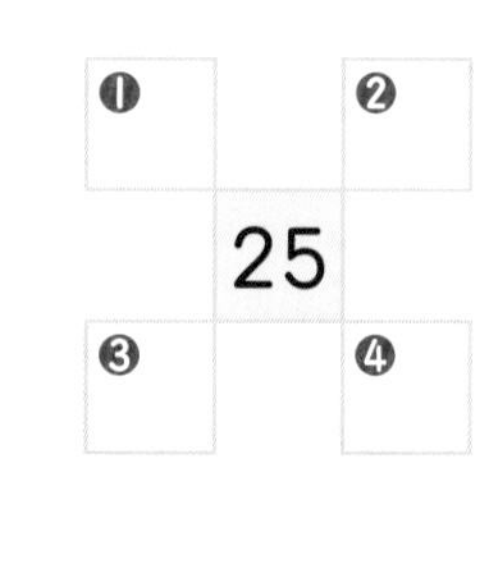

2 There are 3 boxes of chocolates that contain 10 chocolates each, and 10 boxes that contain 6 chocolates each. How many chocolates are there in total?

3 Let's make a math problem for 10×4.
Let's make a math problem for 6×0.

4 Let's find out the numbers of ● using multiplication. Let's write the math expression for it.

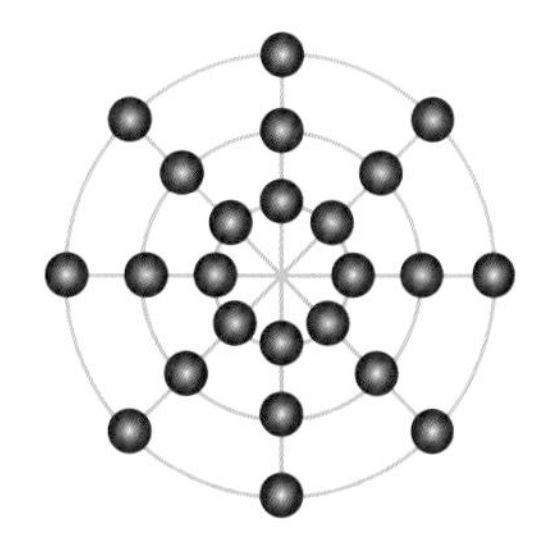

5 Let's think about how to calculate 4×12 using the rules of multiplication.

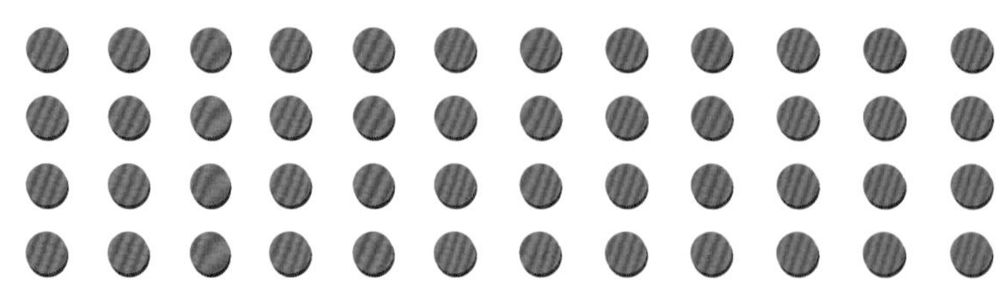

With the Way to See and Think Monsters...

Let's Reflect!

Let's reflect on which monster you used while learning "1 Multiplication."

Rule

Based on the rules of multiplication, we could find out the answers in various ways.

① Let's review the way to find out the answer to 8×10 based on the rules of multiplication we have learned so far.

[Rule of Commutativity]

$8 \times 10 = 10 \times$ ☐ , so ☐ sets of 10 make ☐ .

Sara

[Rule of the Multiplier and the Answer]

In the row of 8, if the multiplier increases by 1, the answer increases by the multiplicand.

$8 \times 10 = 8 \times 9 +$ ☐ , so ☐ .

Akari

[Rule of Distribution]

8×10 < $8 \times 5 =$ ☐ ; $8 \times$ ☐ $=$ ☐

Total ☐

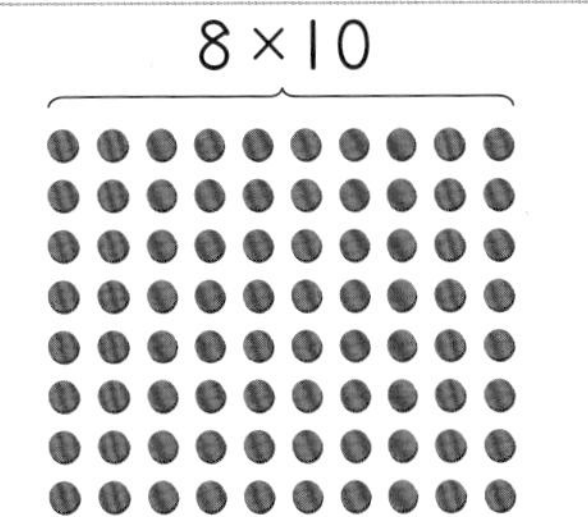

Haruto

[Rule of Associativity]

$8 \times 10 = 8 \times (5 \times 2)$, so

$(8 \times 5) \times 2 =$ ☐ $\times 2$

$40 + 40 =$ ☐

Yu

? Solve the ?

Even if I forgot the answers for the multiplication table, I could find out by using the rules of multiplication.

Haruto

Want to Connect

Can we apply this to a multiplication using larger numbers?

Sara

Find the ?

What time will you arrive?

We are going to go to the aquarium for our field trip.
I want to see the dolphin show.
Let's make a plan for the trip!
1

I got some information about the aquarium. The aquarium opens at 9 a.m. and the dolphin show starts at 9:30 a.m.
2

We will leave school at 8:40 a.m. It will take 30 minutes walk to the aquarium.
3

So can we be on time for the dolphin show?
4

What time will you arrive at the aquarium?

Time and Duration (I)

Let's find out time and duration and utilize in life.

1 Finding out time and duration

What time past ↓

The time they left school

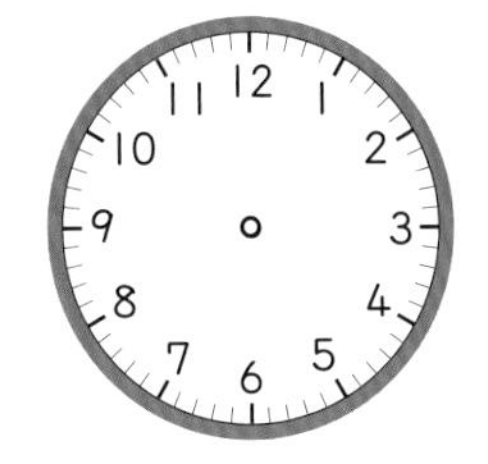

The time they arrived at the aquarium

1 Akari and her friends left school at 8:40 a.m. It took 30 minutes to arrive at the aquarium on foot. What time did they arrive at the aquarium?

They walked for 30 minutes from 8:40 a.m. so we can operate 40 + 30, right?

Sara

That would be 8:70 a.m. ...

Yu

\ Want to know /

Purpose How can we find out the time and duration?

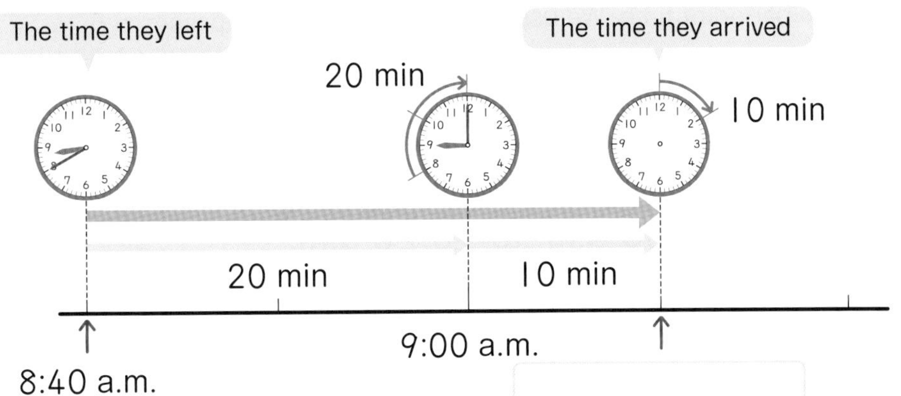

Way to see and think

It is easy to understand by replacing duration or time by a line of number.

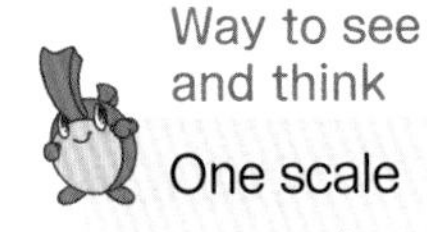

Way to see and think

One scale represents 10 minutes by replacing a clock by a line.

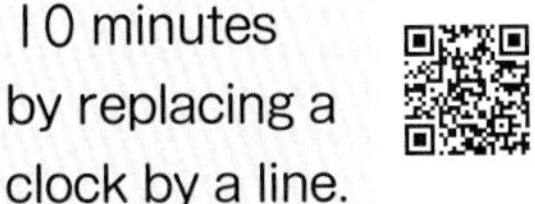

1 Let's find out the following time.

① the time 50 minutes past 9:30 a.m.

② the time 20 minutes past 1:50 a.m.

③ the time 1 hour 10 minutes past 2:40 p.m.

2 They left the aquarium at 9:50 a.m. and arrived at the park at 10:15 a.m. How long did it take from the aquarium to the park?

The time they left the aquarium

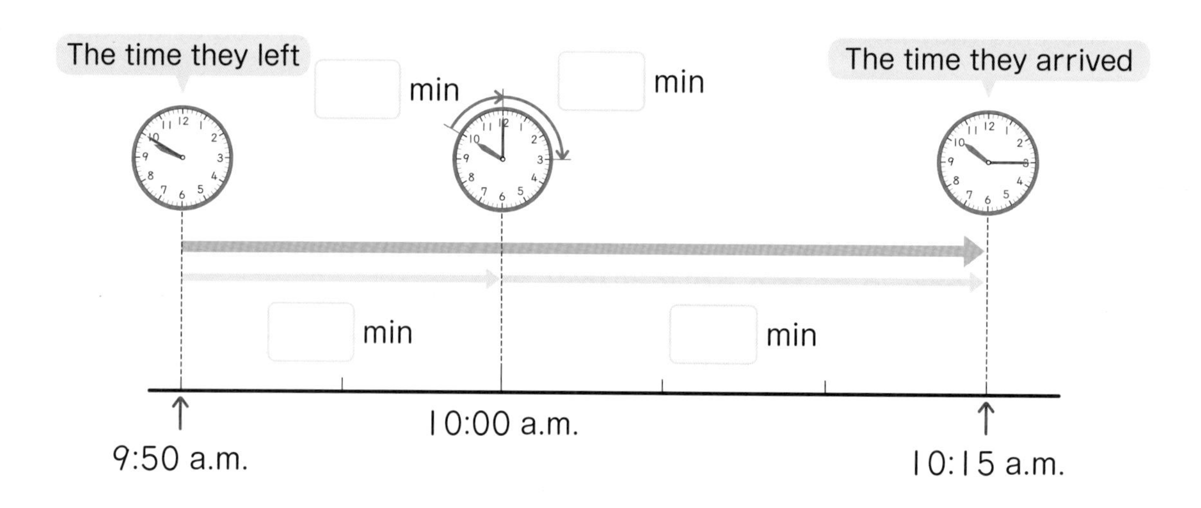

The question of "how long" is answered by duration or period of time. We use "minutes" to represent it.

Way to see and think

One scale represents 5 minutes by replacing a clock by a line of numbers.

3 Let's find out the following time.

① the duration from 9:30 a.m. to 10:20 a.m.

② the duration from 3:10 p.m. to 4:50 p.m.

The time they left the park

The time they arrived at school

2 They left the park and walked for 40 minutes. They arrived at school at 11:30 a.m. What time did they leave the park?

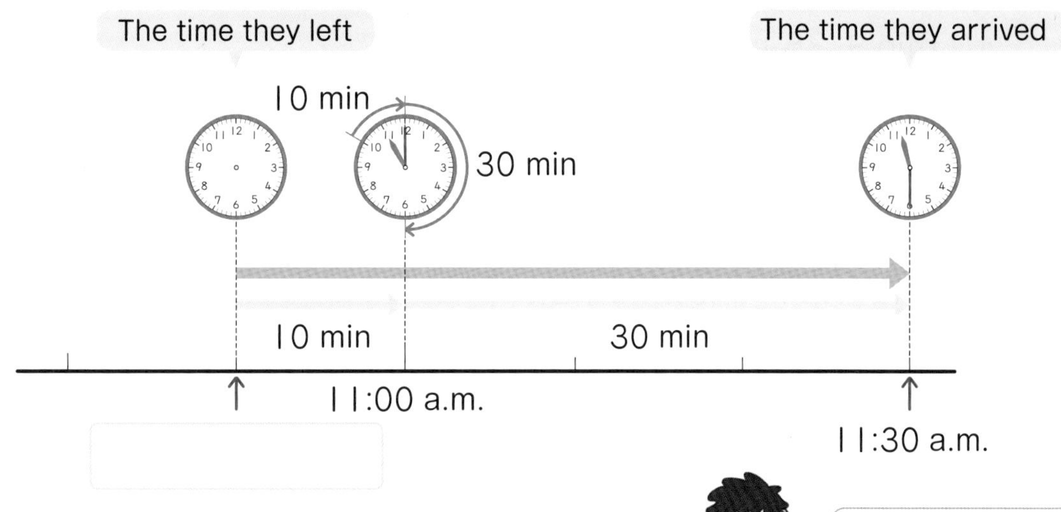

30 minutes before 11:30 a.m. is...
Haruto

1 Let's find out the following time.

① the time 50 minutes before 11:20 a.m.

② the time 1 hour 20 minutes before 2:20 p.m.

Summary

We can find out the time and duration by thinking about the movement of the hands of the clock and replacing them with a line of numbers.

2 They were at the aquarium for 40 minutes and in the park for 35 minutes. How long was the sum of the duration they were in the aquarium and in the park?

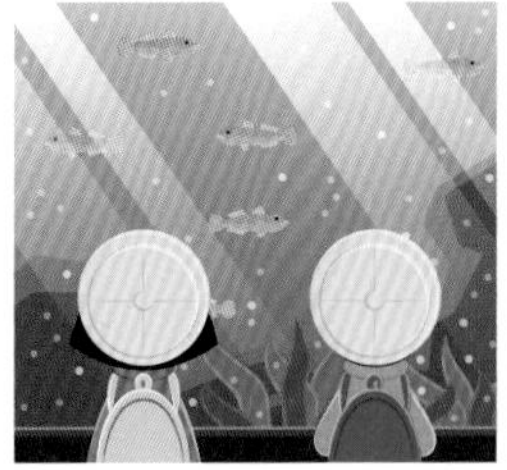

Duration they were in the aquarium 40 min | Duration they were in the park 35 min

0 min | 1 hr

40 min + 35 min = 75 min

75 min = ☐ hr ☐ min

When the minutes become 60, you have to convert and carry 1 to the hours.

3 The duration they walked on their field trip is shown on the right. How long did they walk altogether?

from our school to the aquarium ···30 min
from the aquarium to the park·········25 min
from the park to our school ············40 min

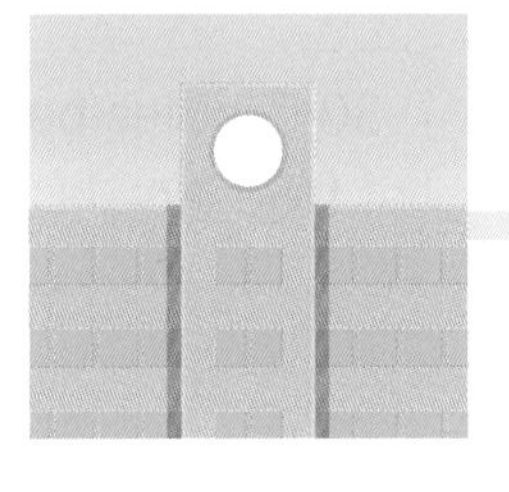

4 They left school at 8:40 a.m. and came back to school at 11:30 a.m. Let's find out the duration they were out of school by writing a math expression.

11:30 ☐ 8:40

Can we calculate time by using the vertical form we have learned in the 2nd grade?

3 Let's think about how to calculate the expression 11:30 − 8:40 you found out in 4 on the previous page by using the vertical form shown on the right.

	hr	min
	11	30
−	8	40

\ Want to think /

Purpose How can we do the carrying and the borrowing?

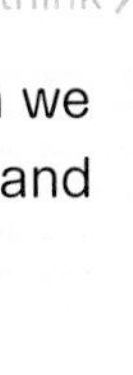

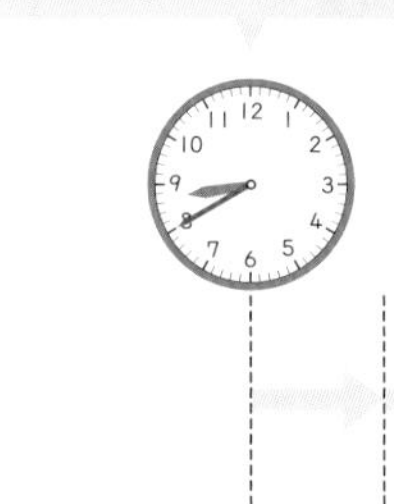

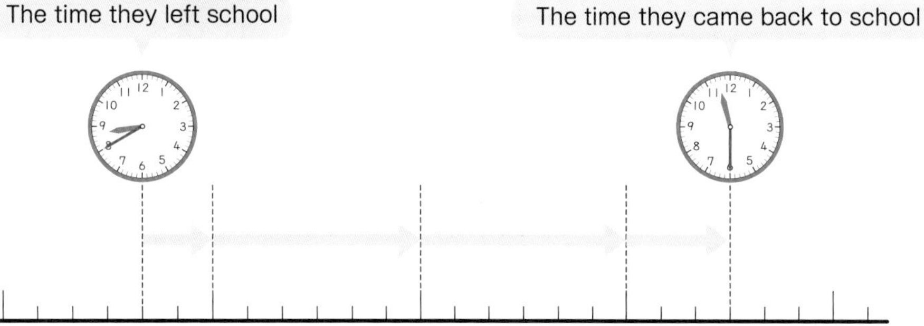

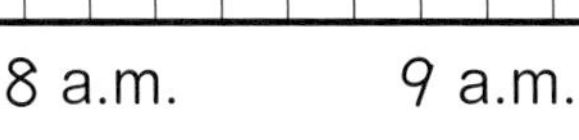

How to calculate time and duration in vertical form ▷

	11	30
−	8	40

Align the "hours" and "minutes."

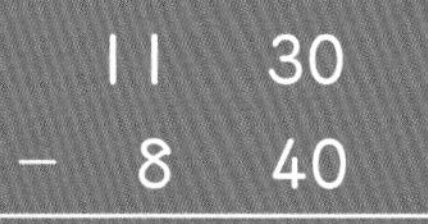

	10	60
	~~11~~	30
−	8	40
		50

1 hour is converted to 60 minutes and borrowed to the minutes.
$60 + 30 = 90$
$90 - 40 = 50$, So, 50 minutes.

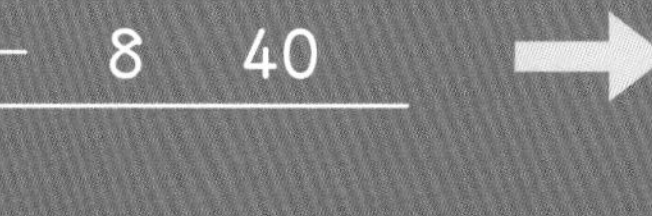

	10	60
	~~11~~	30
−	8	40
	2	50

Because 1 hour is borrowed,
$10 - 8 = 2$
So, 2 hours 50 minutes.

1 What time is 2 hours 50 minutes past 8:30 a.m? Let's calculate using vertical form.

	8	30
+	2	50

When the minutes become 60, they are converted to 1 hour and carried to the hours.

Summary We can think about "hours" and "minutes" as places, and do the carrying and borrowing.

2 Short durations

Seconds ↓

1 Let's examine how long you can stand still on one leg with your eyes closed.

I think I can only stand for shorter than 1 minute...

How can we measure?

\ Want to represent /

? **Purpose** How can we represent duration that are shorter than 1 minute?

A **second** is a unit of time that is shorter than 1 minute. The word second can be written as **sec**.

1 minute = 60 seconds

The second hand takes one second to move from one mark to the next. The second hand makes one rotation in 60 seconds.

Way to see and think

It looks similar to the relationship between 1 hour = 60 minutes.

1 Stopwatches are useful to measure short durations. What does the stopwatch on the right represent?

! **Summary**

We can use seconds to represent durations that are shorter than 1 minute.

2 Let's play the game "Exactly 10 seconds" using a stopwatch. Close your eyes and try to stop the stopwatch when you think 10 seconds have passed.

I want try with 30 seconds and 1 minute too!

? Can we calculate seconds?

2

The table on the right shows the duration of how long the group members of Mirei could stand still on one foot with their eyes closed. Who could stand the longest?

Mirei	58 sec
Yugo	1 min 40 sec
Aina	1 min 28 sec
Fumito	104 sec

Want to compare

Purpose Can we compare when the seconds and the minutes are mixed up?

① Let's think about how to represent in seconds.

Yugo 1 min 40 sec = ☐ sec

Aina 1 min 28 sec = ☐ sec

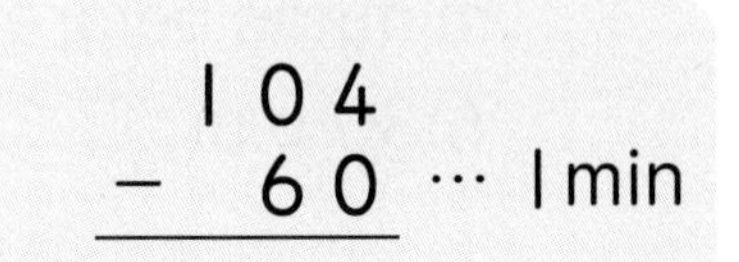

② Let's think about how to represent in minutes and seconds.

Fumito 104 sec = ☐ min ☐ sec

$$\begin{array}{r} 104 \\ -\ 60 \end{array} \cdots 1\text{ min}$$

1 Let's compare the total duration of the time among groups. Let's calculate the total duration for Mirei's group in 2.

Sara's idea: I will calculate by converting all duration into seconds.

Haruto's idea: I will calculate by converting all time into minutes and seconds.

Way to see and think: Can we calculate "minute" and "second" like we did in "hour" and "minute?"

Summary

Way to see and think

Even when the minutes and seconds are mixed up, we can calculate by converting them into one unit.

C A N What can you do?

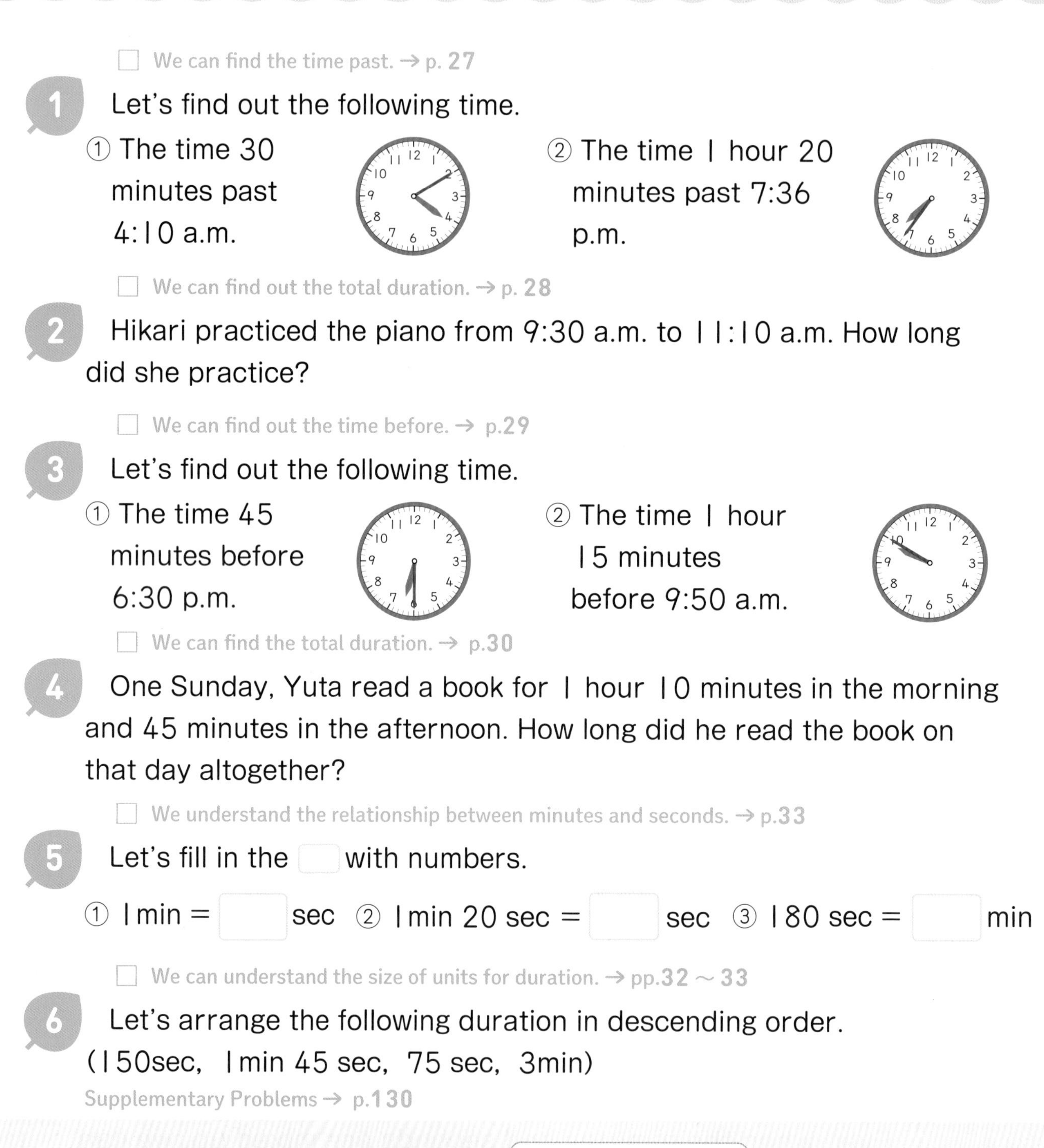

☐ We can find the time past. → p. 27

1 Let's find out the following time.

① The time 30 minutes past 4:10 a.m.

② The time 1 hour 20 minutes past 7:36 p.m.

☐ We can find out the total duration. → p. 28

2 Hikari practiced the piano from 9:30 a.m. to 11:10 a.m. How long did she practice?

☐ We can find out the time before. → p.29

3 Let's find out the following time.

① The time 45 minutes before 6:30 p.m.

② The time 1 hour 15 minutes before 9:50 a.m.

☐ We can find the total duration. → p.30

4 One Sunday, Yuta read a book for 1 hour 10 minutes in the morning and 45 minutes in the afternoon. How long did he read the book on that day altogether?

☐ We understand the relationship between minutes and seconds. → p.33

5 Let's fill in the ☐ with numbers.

① 1 min = ☐ sec ② 1 min 20 sec = ☐ sec ③ 180 sec = ☐ min

☐ We can understand the size of units for duration. → pp.32 ~ 33

6 Let's arrange the following duration in descending order.
(150sec, 1min 45 sec, 75 sec, 3min)

Supplementary Problems → p.130

Which "Way to See and Think Monsters" did you find in "2 Time and Duration (1)"?

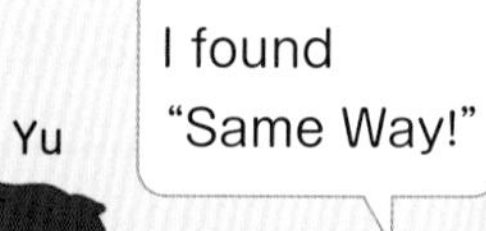

I found "Other Way" when I was trying to find out the time and duration.

Yu

I found "Same Way!"

Sara

With the Way to See and Think Monsters...

Let's Reflect!

Let's reflect on which monster you used while learning "2 Time and Duration (1)."

Other Way

We can find out the time past and duration by representing time by the diagram of the clock or with the line of numbers.

① How did you find out the time 30 minutes past 8:40 a.m.?

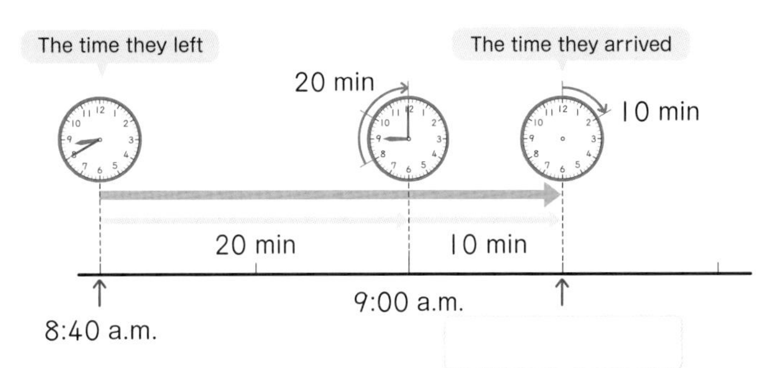

Akari: I used the diagram of the clock. By setting one mark as 1 minute, I moved 30 marks.

Sara: I used the line of numbers. By setting 1 scale as one unit, we can easily draw the line of numbers.

Same Way

Same as the relationship of 1 hour = 60 minutes, we found out that 1 minute = 60 seconds.

② Summarize the relationships among hours, minutes, and seconds.

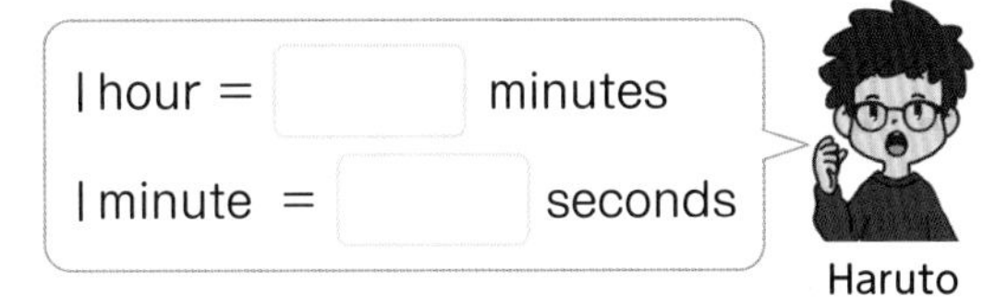

Let's deepen. → p.140

? Solve the ?

We could find out the time past and duration by representing time by the diagram of the clock or with the line of numbers.

Akari

Want to Connect

Can we make a various plans if we know the time and duration?

Haruto

Can we share the cookies equally?

We have 12 cookies. Let's divide them.
Among how many people?
How many each?
1
There are 12 cookies. Let's share them among 4 children.
What is the best way to share the cookies equally?
2
There are 12 cookies. Let's share 4 cookies to each child.
If each child gets the same number of the cookies, how many children can get them?
3
What is the difference between dividing among 4 people and dividing 4 each?
Looks similar, but...
4

What is the difference between "dividing 12 cookies to 4 children equally" and "distributing 4 cookies to each child"?

3 Division

Let's think about how to divide things equally.

1 Calculation to find out the number for each child

1 If you divide 12 cookies equally among 4 children, how many cookies does each child get?

① Let's examine how many cookies each child gets by using blocks.

In this way, the number of the cookies is not equal.

Sara

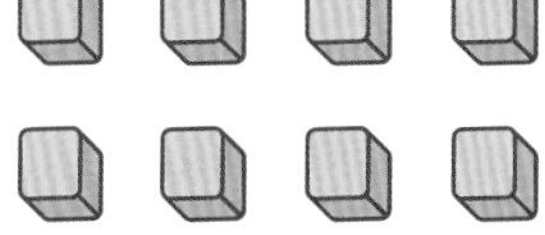
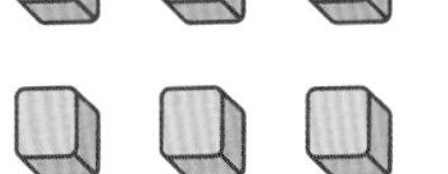

Shall we put 1 on each dish?

Yu

I think we can put 2.

Haruto

Let's put 1 each first.

Akari

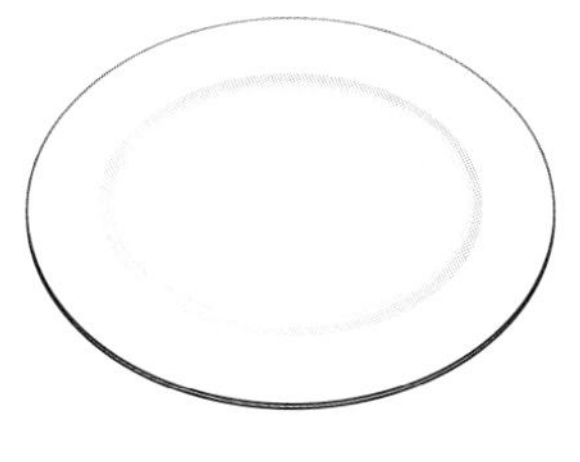
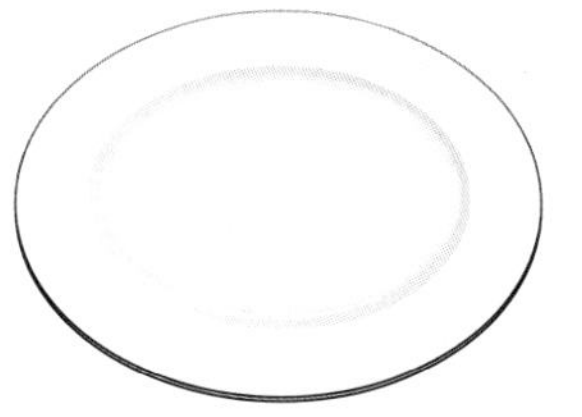

\ Want to think /

? **Purpose** If we divide the cookies equally, how many cookies does each child get?

② Let's explain how to divide the cookies as shown below.

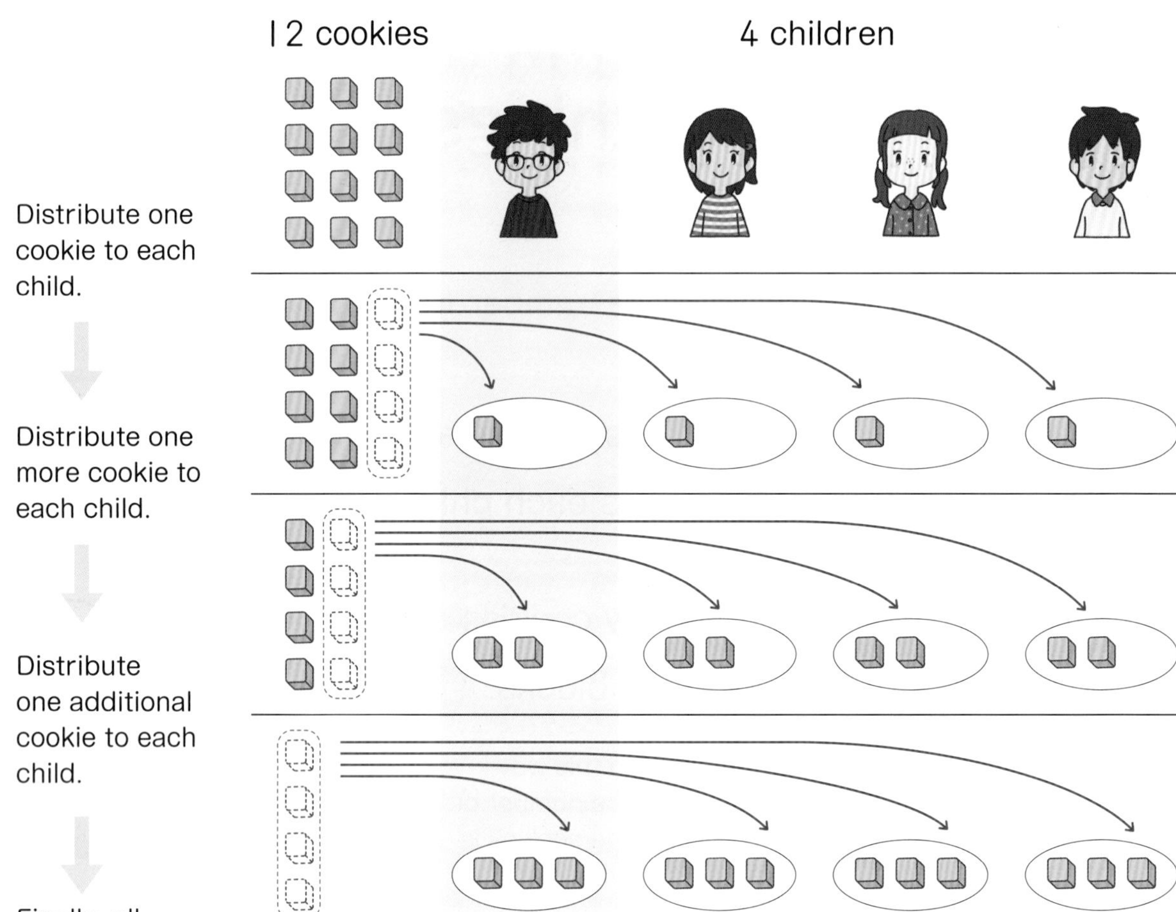

12 cookies
4 children
Distribute one cookie to each child.
Distribute one more cookie to each child.
Distribute one additional cookie to each child.
Finally, all cookies were distributed.

The number of cookies for each child is ☐.

If you divide the 12 cookies equally among 4 children, each child gets 3. In a math sentence, it can be written as $12 \div 4 = 3$, and is read as "12 divided by 4 equals 3."

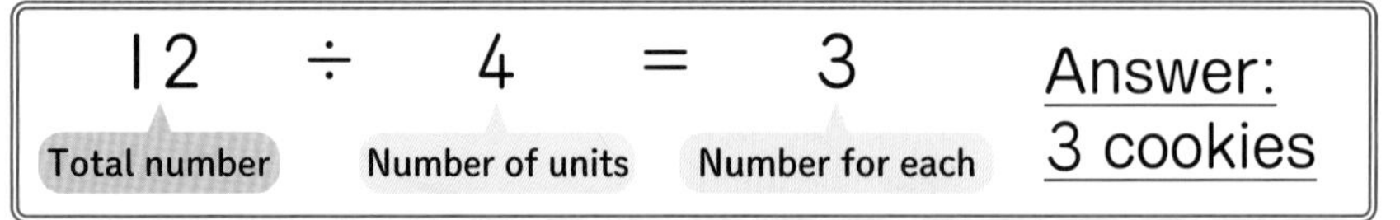

This kind of operation is called **division**.

In this math sentence, Number of units is the number of children, and Number for each is the number of cookies for each child.

1 Let's write a math sentence for each situation below and find out the number of blocks given to each child.

① Divide 6 blocks equally among 3 children.

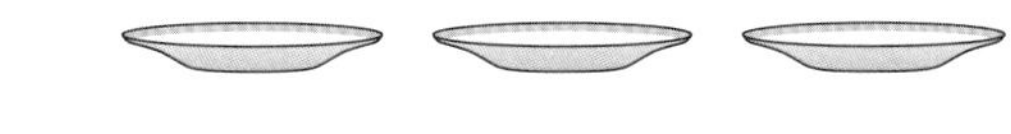

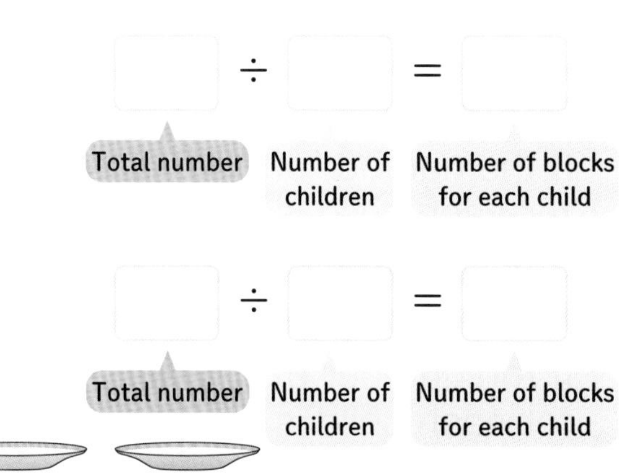

② Divide 15 blocks equally among 5 children.

③ Let's try changing the number of blocks and children.

Summary

When the total number of blocks is equally divided by the number of children, the number of blocks for each child is represented by a division sentence.

Total number ÷ Number of units = Number for each

? Can we find out the answer without using blocks?

2

Divide 15 candies equally among 3 children. How many candies does each child get?

❶ Let's write a math expression.

÷

Total number　Number of units

❷ Let's think about how to calculate.

Yu: How can I find out the number of candies for each child?

Number for each × Number of units = Total number

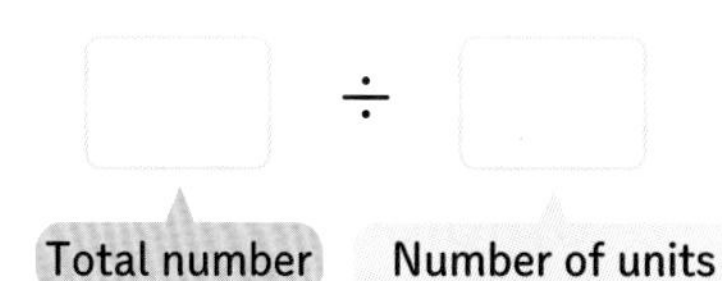

Sara

\ Want to think /

Let's think about how to find out the answer of a division to find out the number of candies for each child.

3 children

		Number for each unit		Number of units		Total number
If the number of candies for each child is	1	1	×	3	=	3
If the number of candies for each child is	2	2	×	3	=	6
If the number of candies for each child is	3	3	×	3	=	9
If the number of candies for each child is	4	4	×	3	=	12
If the number of candies for each child is	5	5	×	3	=	15

Way to see and think

Various numbers are placed as the number for each child.

The answer to $15 \div 3$ can be placed in the □ for $\square \times 3 = 15$.

Math Sentence: $15 \div 3 = 5$

Answer: 5 candies

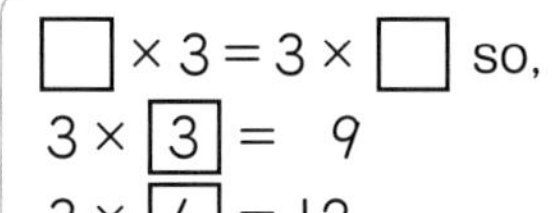

$\square \times 3 = 3 \times \square$ so,

$3 \times \boxed{3} = 9$

$3 \times \boxed{4} = 12$

$3 \times \boxed{5} = (15)$

The answer to $15 \div 3$ can be found by using the row of 3 in the multiplication table.

1 Divide 10 dL of juice equally among 5 children. How many dL of juice does each child get?

$\square \div \square = \square$

Way to see and think

We can think in the same way as we examined $15 \div 3$.

2 Which row of multiplication table should you use to do the following divisions? Let's find out the answers.

① $8 \div 2$ ② $21 \div 7$ ③ $72 \div 9$ ④ $28 \div 4$
⑤ $20 \div 5$ ⑥ $56 \div 8$ ⑦ $21 \div 3$ ⑧ $54 \div 6$

? In what situation do we use division?

Making math problems →

3

Look at the picture and make a problem that can be solved by division.

❶

Akari's idea

☐ chocolates are to be divided equally among ☐ children. How many pieces will each child get?

Want to think

Purpose What do we need to find out in this problem?

Yu

❷

1 Let's make a math problem for the math expression $6 \div 2$.

Haruto: There are 6 apples. If 2 children share the apples equally...

Sara: There are 6 sweets and 2 plates...

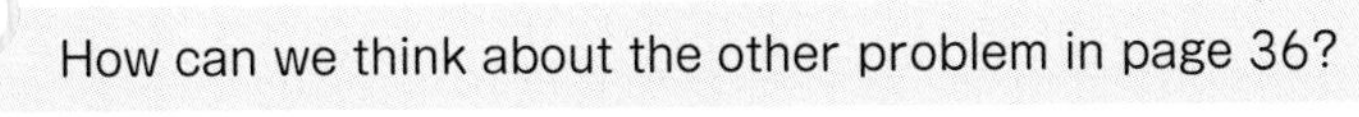

? How can we think about the other problem in page 36?

2 Calculation to find out the number of children

It's the other problem written in "Find the ?" on page 36.

There are 12 cookies. If each child gets 4 cookies, how many children can share the cookies?

① Let's examine how many children can share the cookies by using blocks.

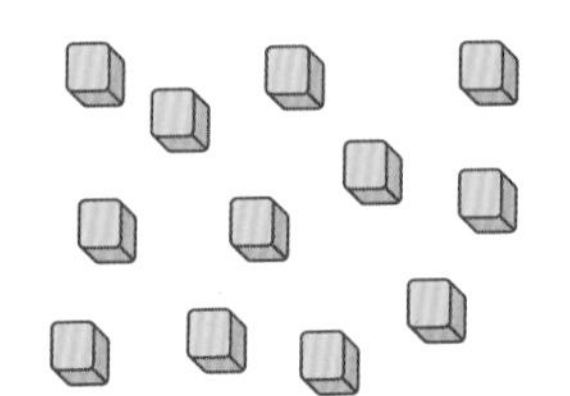

\ Want to think /

Purpose When the cookies are distributed so that each child gets an equal number, how many children can share them?

② Let's explain how to divide the cookies as shown below.

Distribute 4 cookies to one child.

Distribute 4 cookies to one more child.

Distribute 4 cookies to one additional child.

Finally, all cookies were distributed.

☐ children can share the cookies.

If you divide 12 cookies so that each child gets 4 cookies, then 3 children can equally share the cookies. This can also be written by using a division sentence, $12 \div 4 = 3$.

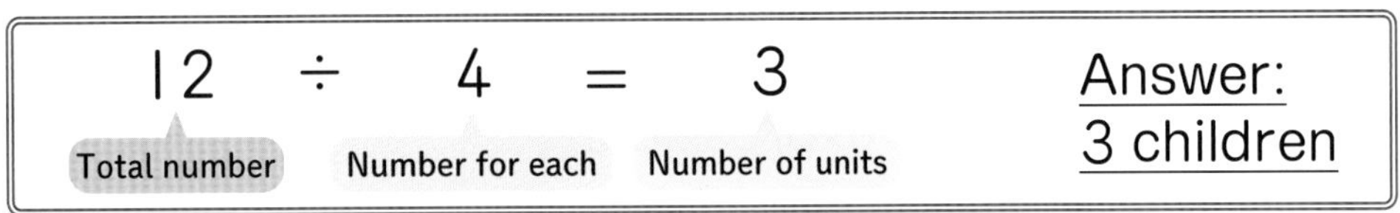

12 ÷ 4 = 3
Total number
Number for each
Number of units
Answer: 3 children

1 Let's write a math sentence for each situation below and find out the number of children who can share the blocks.

① Divide 9 blocks so that each child gets 3 blocks.

$\square \div \square = \square$

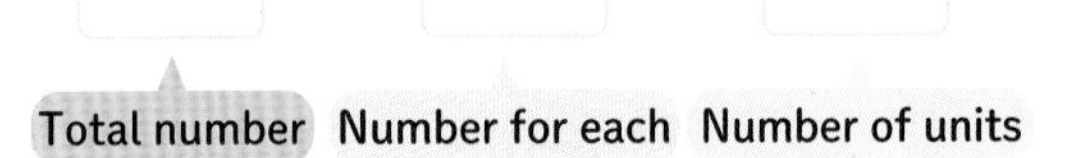

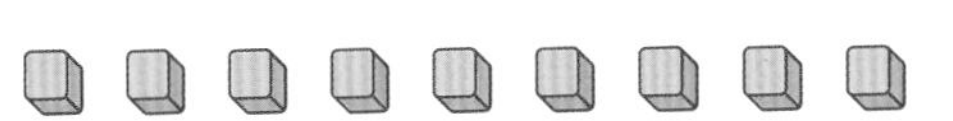

② Divide 20 blocks so that each child gets 4 blocks.

$\square \div \square = \square$

Total number　Number for each　Number of units

③ Let's try changing the number of blocks and for each child.

Summary

When the total number of blocks is distributed so that each child gets an equal number, the number of children who can share the blocks is also represented by a division sentence.

Total number ÷ Number for each = Number of units

2 There are 8 oranges. If you distribute 2 to each child, how many children can share the oranges?

$\square \div \square = \square$

Total number　Number for each　Number of units

? Can we also use the multiplication table to find out the number of children?

2

If you divide 15 candies so that each child gets 3 candies, how many children can share the candies?

1. Let's write a math expression.
2. Let's think about how to calculate.

\ Want to think /

Let's think about how to find out the answer of a division to find out the number of children.

		Number for each		Number of units		Total number
For 1 child	▭	3	×	1	=	3
For 2 children	▭ ▭	3	×	2	=	6
For 3 children	▭ ▭ ▭	3	×	3	=	9
For 4 children	▭ ▭ ▭ ▭	3	×	4	=	12
For 5 children	▭ ▭ ▭ ▭ ▭	3	×	5	=	15

The answer to $15 \div 3$ is the number to be placed in □ for $3 \times □ = 15$.

Math Sentence: $15 \div 3 = 5$ Answer: 5 children

Way to see and think

Various numbers are placed as the number of children.

Summary

The answer to $15 \div 3$ can be found out by using the row of 3 in the multiplication table.

$15 \div 3 = □$

$3 \times \boxed{3} = 9$

$3 \times \boxed{4} = 12$

$3 \times \boxed{5} = 15$ (circled)

1 There are 30 dL of milk. If you drink 6 dL a day, for how many days can you drink the milk?

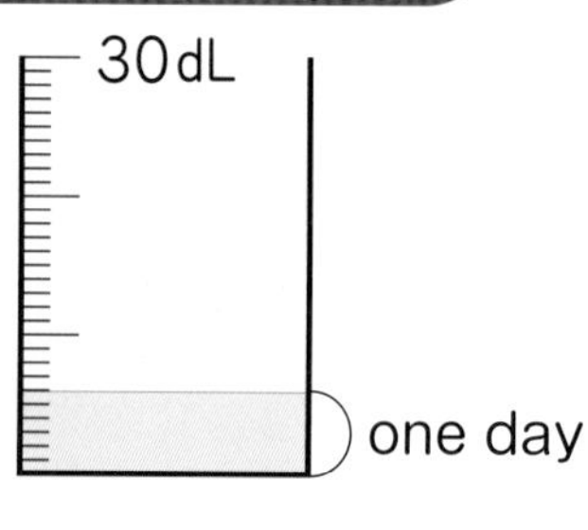

2 Let's divide the following.

① $14 \div 2$ ② $27 \div 9$ ③ $40 \div 5$ ④ $32 \div 8$

⑤ $18 \div 3$ ⑥ $42 \div 7$ ⑦ $48 \div 8$ ⑧ $12 \div 6$

? What is the difference between finding out the number of children and the number for each child?

3 Let's compare the following two problems.

Ⓐ Divide 15 candies equally to 3 children. How many candies will each child get?

Ⓑ There are 15 candies. If we distribute 3 candies to each child, how many children can share the candies?

1 Let's write a math expression for each problems.

Ⓐ

Ⓑ

\ Want to compare /

? **Purpose** Both can be represented in using division expression. What is the difference?

【 Division to find out the number for each child 】

Total number ÷ Number of units = Number for each

【 How to find out the answer 】

5 × 3 = 15

Number for each / Number of units / Total number

【 Division to find out the number of children 】

Total number ÷ Number for each = Number of units

【 How to find out the answer 】

3 × 5 = 15

Number for each / Number of units / Total number

Both are using the multiplication table to find out the answer.

Each □ represents different things.

To find out the number for each child, we used □ × 3 = 15, to find out the number of children, we used 3 × □ = 15.

Summary

Division is an operation to find out the number when we do not know the “Number for each” or “Number of units” in the following multiple sentence:
“Number for each × Number of units = Total number.”

1 Akari and Haruto looked at the picture on the right and made problems for $10 \div 5$ as below. Let’s compare the differences between them by expressing it in diagrams or expressions.

Problem

Akari’s problem

Divide 10 tomatoes equally into 5 plates. How many tomatoes will be there on each plate?

Haruto’s problem

There are 10 tomatoes. 5 tomatoes are placed on each plate. How many plates are needed?

① Let’s circle the tomatoes in the diagram below so as to represent each problem.

Diagram

② Let’s find out the answer to each problem.

Math expression

Akari’s method

I thought about the operation to find out the number in □.
“□ × 5 = 10.” So, 10 ÷ 5 = □
Answer: □ tomatoes

Haruto’s method

I thought about the operation to find out the number in □.
“5 × □ = 10”. So, 10 ÷ 5 = □
Answer: □ plates

Even though the expression of division is the same, we are trying to find out different things.

Sara

In the math expression $10 \div 5$, 10 is called the **dividend,** and 5 is called the **divisor**. The answer to a division problem can be found by using the row of the divisor in the multiplication table.

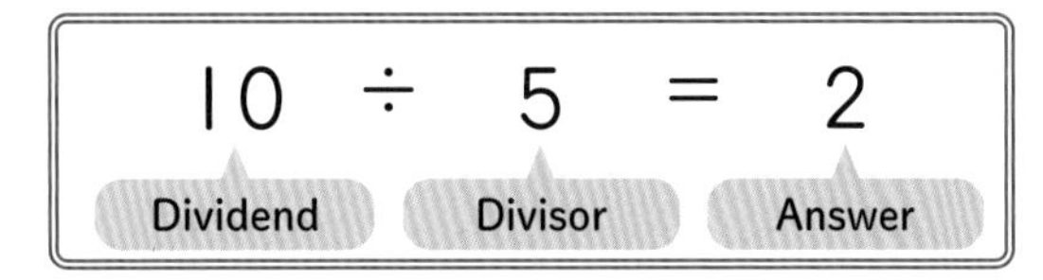

? What can we do if we have 0 or 1 in division?

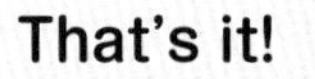

That's it!

Making Books of Division

Let's make two types of books about division based on what we have learned.

① A book of division to find out the number for each.

② A book of division to find out the number of units.

3 Division with 1 and 0

There are cookies in a box to be shared equally by 4 children.
How many cookies does each child get?

① If there are 12 cookies,

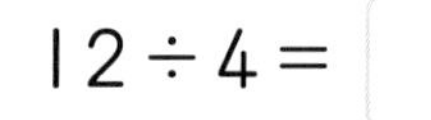

$12 \div 4 = \square$

② If there are 4 cookies,

$4 \div 4 = \square$

③ If there are no cookies,

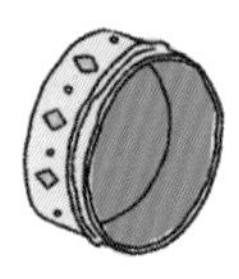

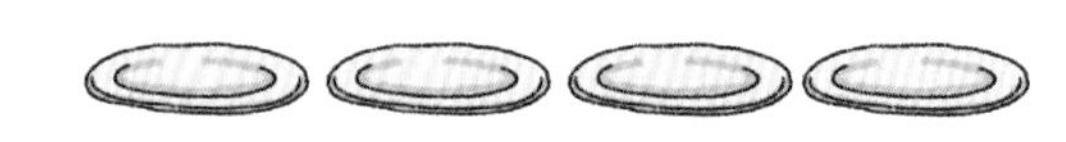

$0 \div 4 = \square$

Want to think

Purpose What should we do if the dividend is 0?

Akari

1 There is 6dL juice to be poured into 1 dL cups.
How many cups are needed?

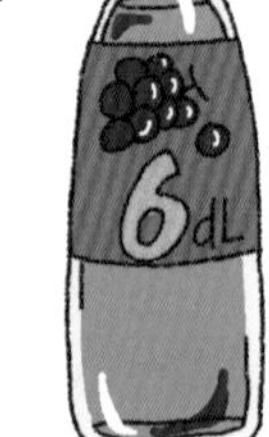

2 Let's divide the following.

① $6 \div 6$ ② $7 \div 7$ ③ $2 \div 2$ ④ $5 \div 5$

⑤ $0 \div 8$ ⑥ $0 \div 3$ ⑦ $5 \div 1$ ⑧ $9 \div 1$

4 Using the rules of operation

1 There are 80 sheets of colored paper that are to be shared equally by 4 children. How many sheets does each child get?

① Let's write a math expression.

I can't find 80 as the answer in the multiplication table...

80 is the number that is the sum of 8 sets of 10.

\ Want to think /

? **Purpose** What should we do when we cannot use the multiplication table?

② Akari and Haruto are thinking about how calculate as shown below. Let's continue to write each idea in your notebook and explain it. Then, let's find out the answer by using each idea.

Akari's idea

$80 \div 4$

80 can be considered as 8 sets of 10.

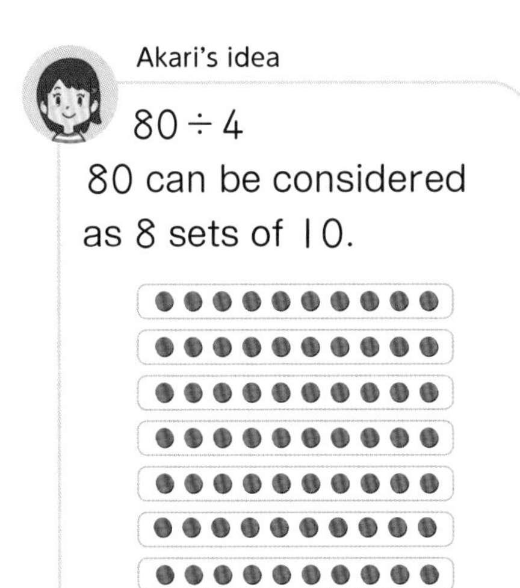

Haruto's idea

$80 \div 4$

The dividend can be divided into 2.

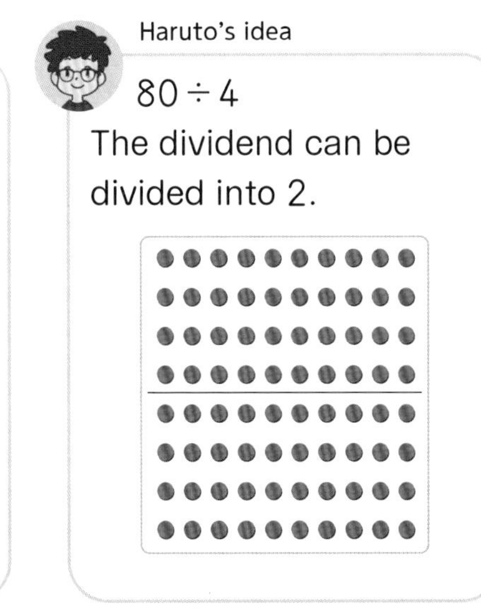

Way to see and think

80 can be considered as 8 sets of 10.

1 Let's calculate the following.

① $90 \div 3$　② $40 \div 4$　③ $80 \div 2$

2 Let's think about how to calculate $36 \div 3$.

① Yu and Sara thought about how to calculate as shown below. Let's explain each idea.

Yu's idea

I thought of the row of 3 in the multiplication table.

$3 \times \boxed{9} = 27$

This is still not enough to make 36. So,

$3 \times \boxed{10} = 30$) +3

$3 \times \boxed{11} = 33$) +3

$3 \times \boxed{12} = 36$

Thus, $36 \div 3 = 12$ Answer: 12

Way to see and think

In multiplication, if the multiplier increases by 1, the answer increases by the multiplicand.

Sara's idea

I thought of decomposing 36 into 30 and 6.

$30 \div 3 = 10$

$6 \div 3 = 2$

Then, $10 + 2 = 12$

Thus, $36 \div 3 = 12$ Answer: 12

Way to see and think

As in multiplication, a dividend can be divided in division.

② By using each idea above, let's explain how to calculate $24 \div 2$ and $39 \div 3$.

! **Summary** Way to see and think

When we cannot use the multiplication table, we can find out the answer for division by considering sets of 10 and by dividing the dividend.

1 Let's calculate the following.

① $48 \div 4$ ② $63 \div 3$ ③ $33 \div 3$

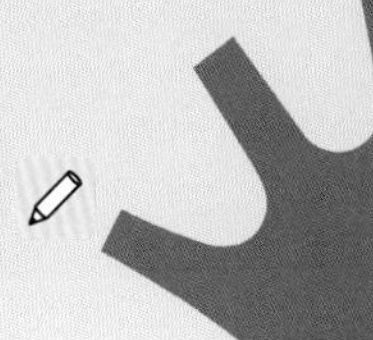

CAN What can you do?

☐ We can find out the answer to division by using the multiplication table. → pp.39 ~ 50

1 Let's calculate the following.

① $27 \div 3$ ② $12 \div 2$ ③ $56 \div 7$ ④ $40 \div 8$

⑤ $63 \div 9$ ⑥ $25 \div 5$ ⑦ $16 \div 4$ ⑧ $49 \div 7$

⑨ $28 \div 7$ ⑩ $54 \div 9$ ⑪ $7 \div 1$ ⑫ $9 \div 9$

⑬ $0 \div 6$ ⑭ $3 \div 3$ ⑮ $0 \div 9$ ⑯ $4 \div 1$

⑰ $70 \div 7$ ⑱ $60 \div 2$ ⑲ $88 \div 4$ ⑳ $96 \div 3$

☐ We can find out the number for each. → p.39

2 Let's answer the following.

① There are 36 strawberries that are to be shared equally by 4 children. How many strawberries does each child get?

② Divide 24 sheets of origami paper equally into 3 bags. How many sheets are in each bag?

☐ We can find out the number of units. → p.42, p.48

3 Let's answer the following.

① There are 24 pencils. When they are divided into 4 pencils each in bundles, how many bundles will be made?

② There is a 8 m ribbon. When it is cut into 1 m pieces, how many pieces are there?

☐ We understand which should be found, the number for each or the number of units. → pp.45 ~ 46

4 There is 14 dL juice. Let's think about the following.

① When each child gets 7 dL, how many children can share the juice?

② Divide the juice equally among 7 children. How many dL does each child get?

Supplementary Problems → p.131

Which "Way to See and Think Monsters" did you find in ," 3 Division"?

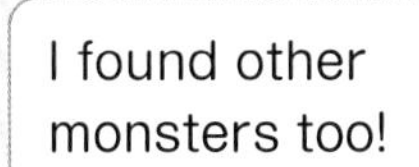

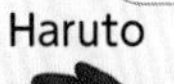

I found "Same Way" when I was comparing the two division problems.

Haruto

I found other monsters too!

Akari

Utilize Usefulness and Efficiency of Learning

1 Let's calculate the following.

① $35 \div 7$	② $72 \div 8$	③ $16 \div 8$
④ $48 \div 6$	⑤ $12 \div 3$	⑥ $45 \div 5$
⑦ $10 \div 2$	⑧ $35 \div 5$	⑨ $64 \div 8$
⑩ $36 \div 6$	⑪ $4 \div 2$	⑫ $16 \div 2$
⑬ $81 \div 9$	⑭ $63 \div 7$	⑮ $42 \div 6$
⑯ $3 \div 1$	⑰ $8 \div 8$	⑱ $0 \div 2$
⑲ $50 \div 5$	⑳ $77 \div 7$	㉑ $84 \div 2$

2 There are 36 colored paper.

① Divide the papers equally among 9 children. How many papers does each child get?

② When each child gets 9 papers, how many children can share the papers?

3 Make a math problem for $32 \div 4$.

Let's fill in the ☐ with numbers or words.

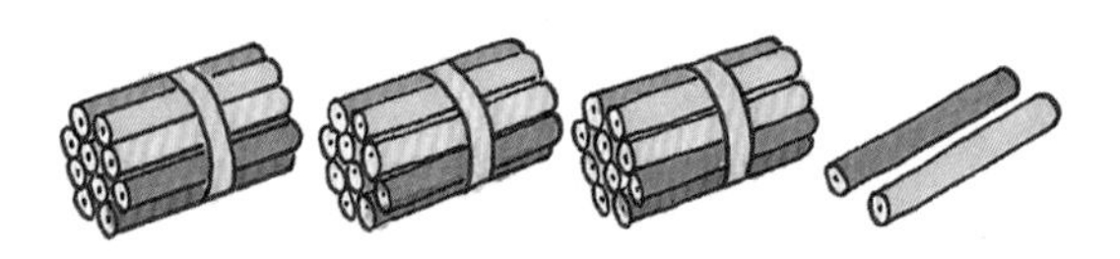

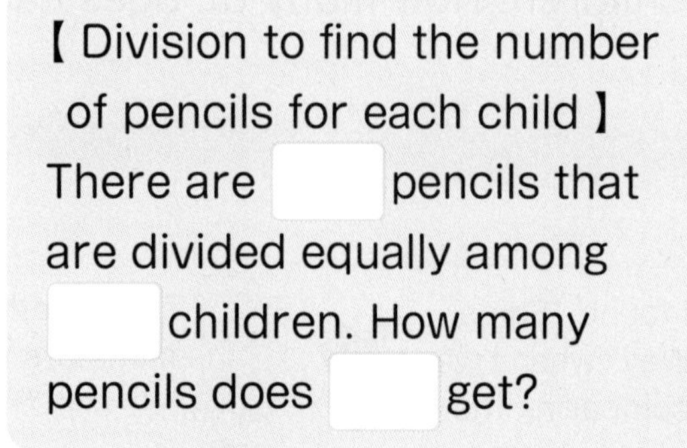

【 Division to find the number of pencils for each child 】

There are ☐ pencils that are divided equally among ☐ children. How many pencils does ☐ get?

【 Division to find the number of children 】

There are ☐ pencils. Each child gets ☐ pencils. How many ☐ can share the pencils?

With the Way to See and Think Monsters...

Let's Reflect!

Let's reflect on which monster you used while learning "3 Division."

Same way

In both cases of division problem so as to find out the number of each and number of units, although what we want to find out is different, we could find out the answer in the same way by applying various numbers or using the multiplication table.

① Let's think about a math problem for division.

Ⓐ Divide 15 candies equally among 3 children. How many candies does each child get?

Ⓑ If you divide 15 candies so that each child gets 3 candies, how many children can share the candies?

We applied various numbers to find out the answer.

Sara

Ⓐ is a problem to find out the number for each, and Ⓑ is to find out the number of units.

Yu

We could find out the answers for both problems by using the multiplication table.

Akari

② Which math sentence corresponds to the math problem Ⓐ and Ⓑ above?

Ⓒ Total number ÷ Number for each = Number of units

Ⓓ Total number ÷ Number of units = Number for each

We can find out the total number by number for each × number of units.

Haruto

Solve the ?

Number of for each and number of units could also be found out by using the division expression.

Yu

Want to Connect

In the problem above, we cannot distribute equally if we have 20 candies to 3 children. What can we do?

Sara

Let's Try!

How many times?

Let's think about how many times by comparing the length of the tapes.

1 Tape Ⓐ is three times long as tape Ⓑ. When the length of tape Ⓑ is 4cm what is the length of tape Ⓐ?

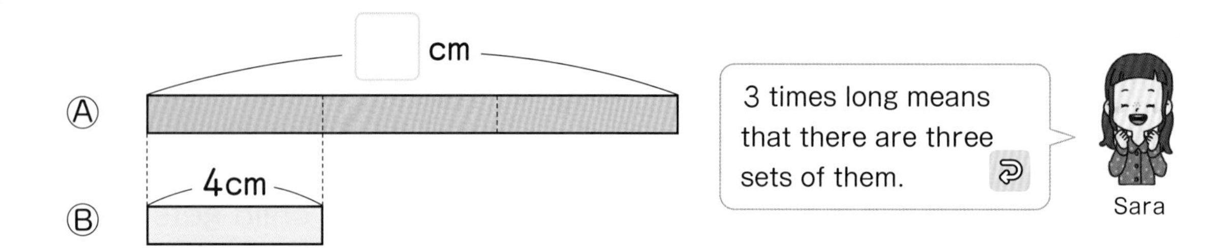

cm
Ⓐ
4cm
Ⓑ
3 times long means that there are three sets of them.
Sara

① Let's write a math sentence and the answer.

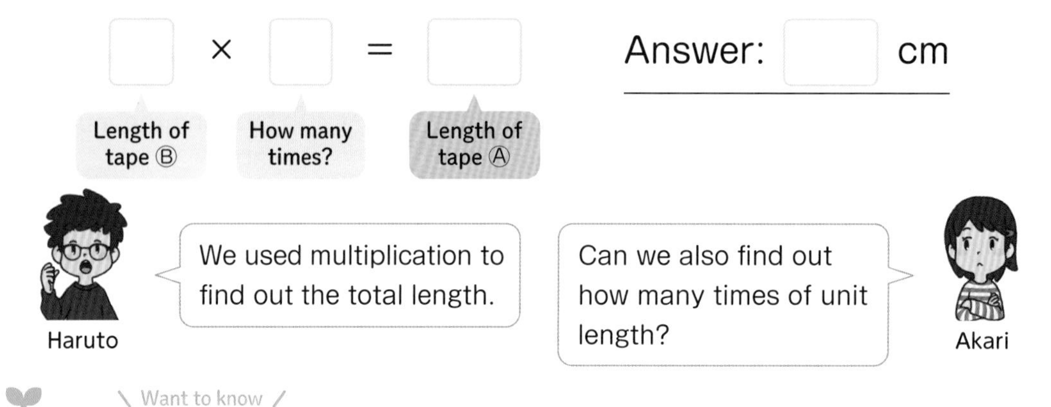

\ Want to know /

Purpose What should we do to find out how many times of unit length?

1 There are 3 tapes as shown below. Let's think about the lengths of these tapes.

Ⓒ 12cm

Ⓓ 6cm

Ⓔ 3cm

① How many tape Ⓓ is needed to make the length of tape Ⓒ ?

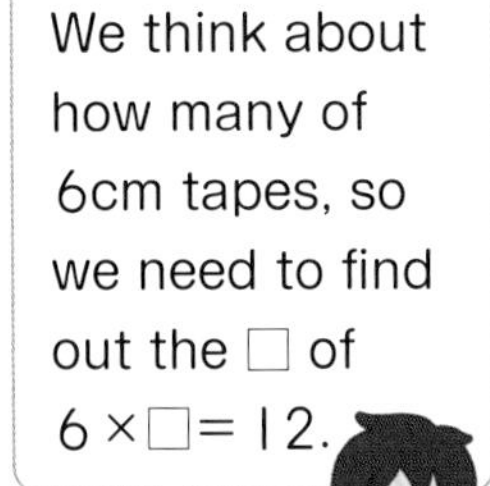

We think about how many of 6cm tapes, so we need to find out the □ of 6 × □ = 12.
Yu

Ⓒ

Ⓓ

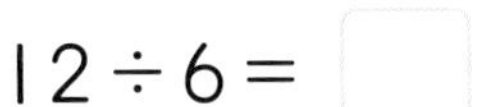

$12 \div 6 =$ □　　Answer: □ tapes

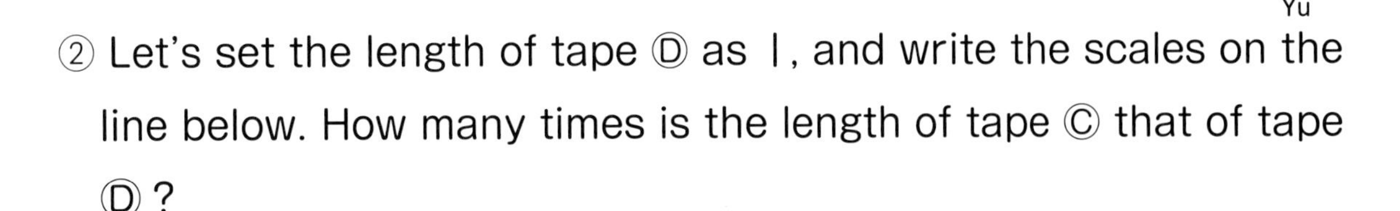

② Let's set the length of tape Ⓓ as 1, and write the scales on the line below. How many times is the length of tape Ⓒ that of tape Ⓓ ?

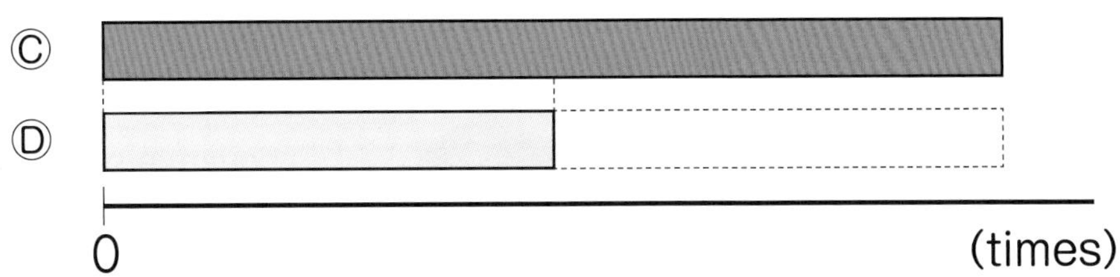

③ How many times is the length of tape Ⓒ that of tape Ⓔ ?

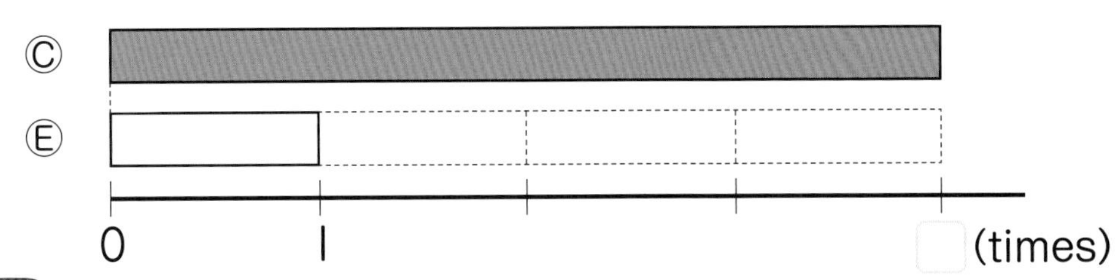

Summary

To find out how many times of unit length, division is used.

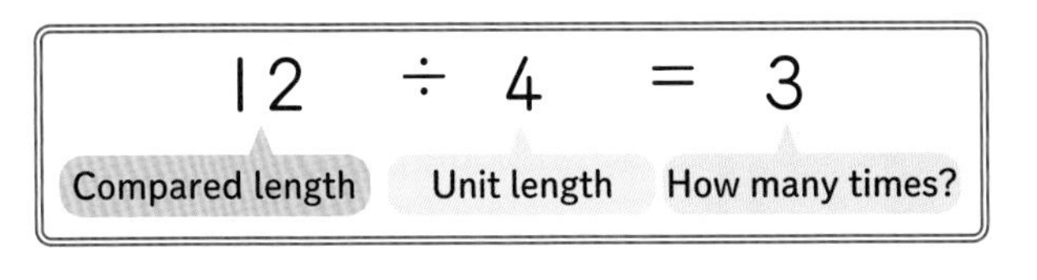

2 The water tank in the science lab holds 24L of water. The water tank in 3rd grade classroom holds 6L of water.
How many times more water does the water tank in the science lab hold than the tank in the 3rd grade classroom?

How much is the total cost?

\ Want to think /

Purpose How can we add 3-digit numbers in vertical form?

Addition and Subtraction

Let's think about how to calculate 3-digit numbers in vertical form.

1 Addition of 3-digit numbers

1 You bought a biscuit for 215 yen and a chocolate for 143 yen. How much was the total cost?

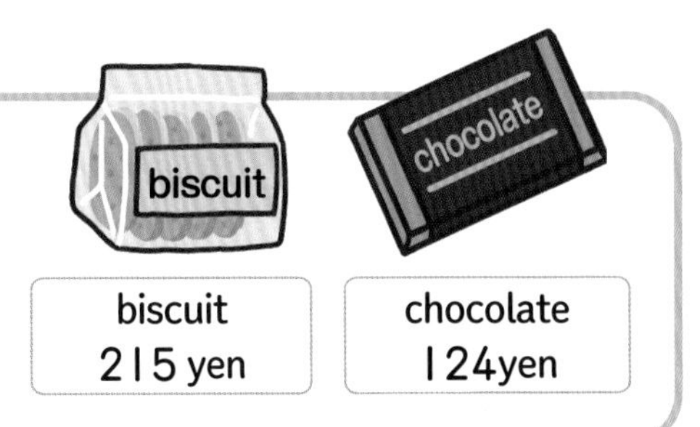

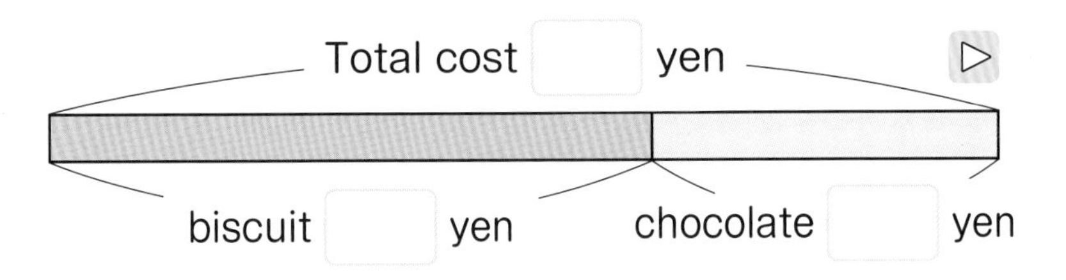

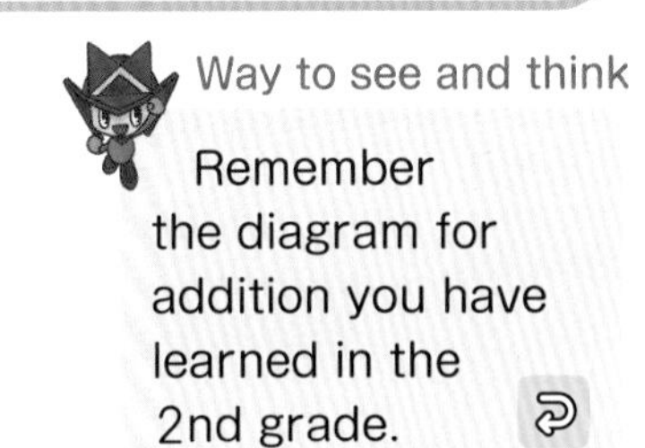

Way to see and think
Remember the diagram for addition you have learned in the 2nd grade.

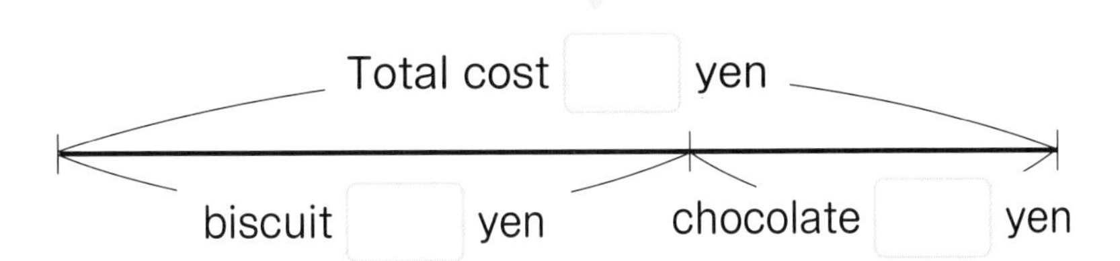

We can draw easily by only using a line.

① Let's write a math expression.

② Approximately, how much is it? yen

300 yen is not enough.

Sara

③ Let's think about how to calculate.

I'll try to use the same idea as in adding 2-digit numbers by using blocks.

Yu

Akari

Can I add in vertical form?

Way to see and think

Can we think of it like addition we have learned in the 2nd grade?

Yu's idea

Align the blocks vertically according to their places and then add the numbers in each place.

Hundreds	Tens	Ones
boxes of 100 2 + 1 = 3	boxes of 10 1 + 2 = 3	single blocks 5 + 4 = 9

215 + 124 = 339

Akari's idea

Use addition in vertical form like what we did in addition of 2-digit numbers.

$$\begin{array}{r} 215 \\ +\ 124 \\ \hline 339 \end{array}$$

In 2nd grade, we calculated vertical form from the ones place.

Addition algorithm for 215 + 124 in vertical form

Align the digits of the numbers according to their places.

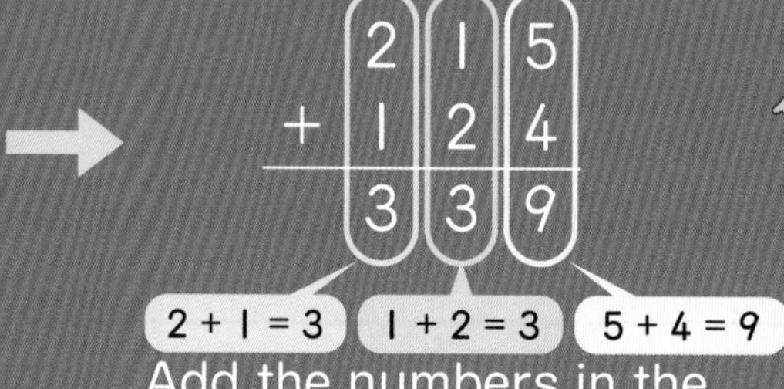

Add the numbers in the same places.

Way to see and think

Even when the numbers get large, we can align the digits of the numbers according to their places.

Summary

For adding 3-digit numbers in vertical form, align the digits of the numbers according to their places and then add the numbers in the same place.

1 Let's calculate the following in vertical form.

① 153 + 425 ② 261 + 637 ③ 437 + 302 ④ 502 + 207

? What should I do if there is carrying?

2

The number of books borrowed from the library was 217 in April and 326 in May. What was the total number of books borrowed in two months?

① Let's write a math expression.

\ Want to think /

Purpose 7+6=13 in the ones place, so we have carrying to the tens place. How can we do?

Haruto

② Let's think about how to add in vertical form.

$$\begin{array}{r} 217 \\ +\,326 \\ \hline \end{array} \rightarrow \begin{array}{r} 217 \\ +\,326 \\ \hline {}^{1}3 \end{array} \rightarrow \begin{array}{r} 217 \\ +\,326 \\ \hline 4{}^{1}3 \end{array} \rightarrow \begin{array}{r} 217 \\ +\,326 \\ \hline 54{}^{1}3 \end{array}$$

1 Let's explain how to calculate 275 + 564 in vertical form.

$$\begin{array}{r} 275 \\ +\,564 \\ \hline \end{array} \rightarrow \begin{array}{r} 275 \\ +\,564 \\ \hline 9 \end{array} \rightarrow \begin{array}{r} 275 \\ +\,564 \\ \hline {}^{1}39 \end{array} \rightarrow \begin{array}{r} 275 \\ +\,564 \\ \hline 8{}^{1}39 \end{array}$$

Don't forget to write down the number that you carried.

Summary For addition of 3-digit numbers, we can carry one to the higher place as we learned.

Sara

2 Let's calculate the following in vertical form.

① 258 + 234 ② 512 + 249 ③ 308 + 415 ④ 102 + 418

⑤ 324 + 195 ⑥ 253 + 574 ⑦ 625 + 190 ⑧ 576 + 62

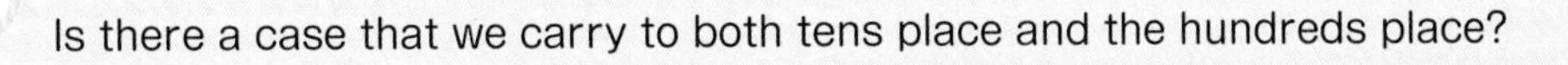

? Is there a case that we carry to both tens place and the hundreds place?

Let's think about how to calculate 248 + 187 in vertical form.

\ Want to think /

Purpose We get numbers larger than 10 for both the ones place and the tens place. What can I do?

Akari

Addition algorithm for 248 + 187 in vertical form

Hundreds	Tens	Ones

③ Hundreds Place
$2+1+1=4$

② Tens Place
$4+8+1=13$
Carry 10 tens to the hundreds place as 1 hundred.

① Ones Place
$8+7=15$
Carry 10 ones to the tens place as 1 ten.

$$\begin{array}{r} 248 \\ +\ 187 \\ \hline \end{array} \;\blacktriangledown\; \begin{array}{r} 248 \\ +\ 187 \\ \hline 5 \end{array} \;\blacktriangledown\; \begin{array}{r} 248 \\ +\ 187 \\ \hline 35 \end{array} \;\blacktriangledown\; \begin{array}{r} 248 \\ +\ 187 \\ \hline 435 \end{array}$$

1 Let's explain how to calculate 346 + 257 in vertical form.

$$\begin{array}{r} 346 \\ +\ 257 \\ \hline \end{array} \;\triangleright\; \begin{array}{r} 346 \\ +\ 257 \\ \hline 3 \end{array} \;\triangleright\; \begin{array}{r} 346 \\ +\ 257 \\ \hline 03 \end{array} \;\triangleright\; \begin{array}{r} 346 \\ +\ 257 \\ \hline 603 \end{array}$$

Summary When we have carrying for both the ones place and the tens place, we can do the carrying as we learned so far.

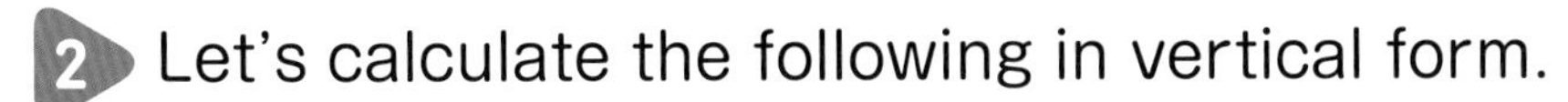

2 Let's calculate the following in vertical form.

① 376 + 546　② 453 + 367　③ 859 + 51

④ 537 + 168　⑤ 456 + 344　⑥ 737 + 68

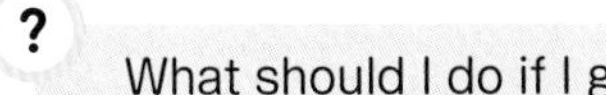

? What should I do if I get numbers larger than 10 when I add the hundreds place?

4 Let's think about how to calculate the following in vertical form.

\ Want to think /

Purpose What should I do if I have carrying in the hundreds place?

Haruto

1 856 + 327

```
   8 5 6        8 5 6        8 5 6        8 5 6
 + 3 2 7  →   + 3 2 7  →   + 3 2 7  →   + 3 2 7
 -------      -------      -------      -------
                  ¹3          8¹3       1 1 8¹3
```

2 459 + 563

```
   4 5 9        4 5 9        4 5 9        4 5 9
 + 5 6 3  →   + 5 6 3  →   + 5 6 3  →   + 5 6 3
 -------      -------      -------      -------
                  ¹2        ¹2¹2        1 0¹2¹2
```

Summary Whichever place we have the carrying, we always carry to the higher place.

1 Let's calculate the following in vertical form.

① 643 + 628　② 747 + 563　③ 526 + 474

④ 888 + 345　⑤ 872 + 129　⑥ 906 + 95

? Can I do 3-digit subtraction in the same way as addition?

Find the ?

How much money is left?

candy
15yen
gummy
48yen
jelly
24yen
gum
27yen
cookie
56yen
biscuit
215yen
chocolate
124yen
What should I buy?
328yen
102yen
80yen
I bought a jelly.
jelly
24yen
80yen
You have 56 yen left.
I bought a gum.
gum
27yen
102yen
You have 75 yen left.
I bought a biscuit.
How much is left?
biscuit
215yen
328yen

\ Want to think /

Purpose How can we subtract 3-digit numbers in vertical form?

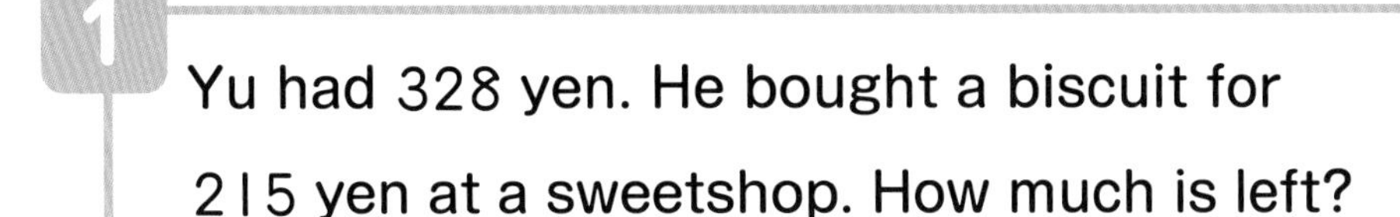

2 Subtraction of 3-digit numbers

1 Yu had 328 yen. He bought a biscuit for 215 yen at a sweetshop. How much is left?

biscuit
215 yen

① Let's draw a diagram.

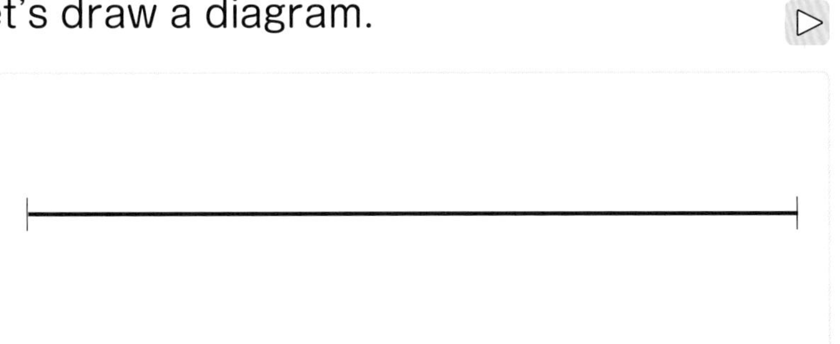

Way to see and think

Remember the diagram for subtraction we have learned in the 2nd grade.

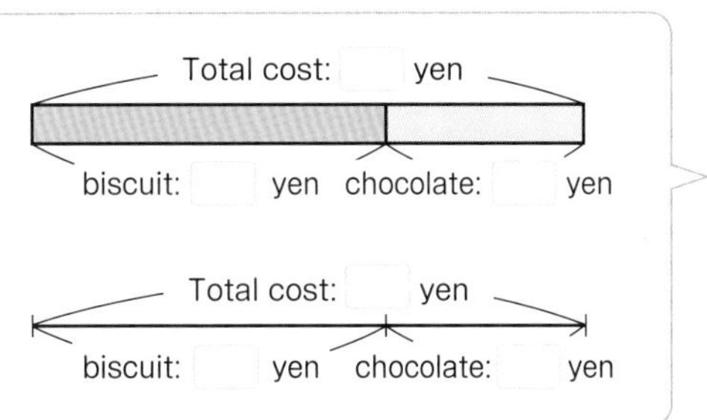

When we learned addition, it was easier to draw a diagram using a line.

② Let's write a math expression.

③ Approximately, how much is left?  yen

Is more than 100 yen left?

Haruto

④ Let's think about how to calculate.

Yu

I'll try to use the same idea as in subtracting 2-digit numbers by using blocks.

Can I subtract in vertical form like I did in addition?

Sara

Way to see and think

Can we think of it like subtraction we have learned in the 2nd grade?

Yu's idea

Represent the number of each places using blocks, and subtract the numbers in each place.

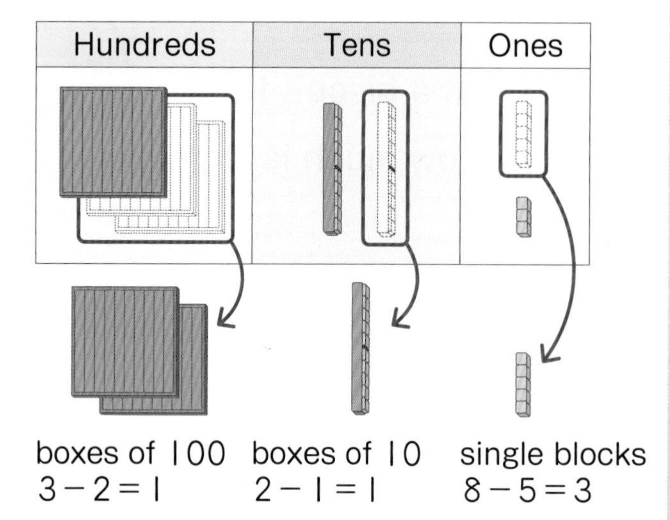

boxes of 100 $3-2=1$ | boxes of 10 $2-1=1$ | single blocks $8-5=3$

$328-215=113$

Sara's idea

Use subtraction in vertical form like what we did in subtraction of 2-digit numbers.

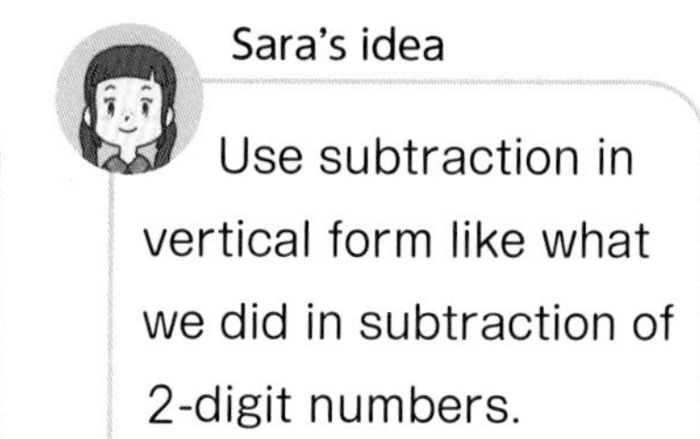

$$\begin{array}{r} 328 \\ -\ 215 \\ \hline 113 \end{array}$$

In 2nd grade, we calculated vertical form from the ones place.

Subtraction algorithm for 328 − 215 in vertical form

$$\begin{array}{r} 328 \\ -\ 215 \\ \hline \end{array}$$

Align the digits of the numbers according to their places.

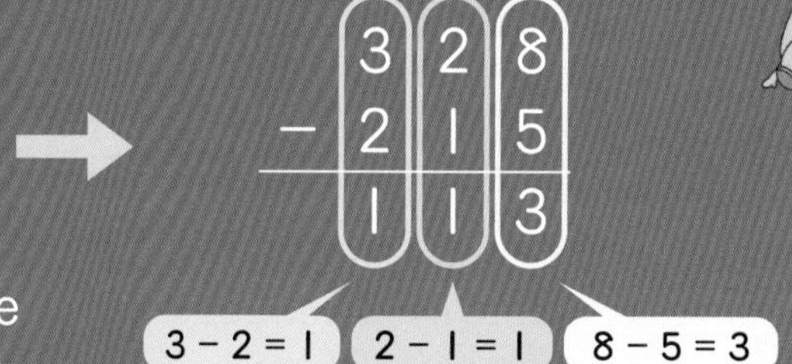

Subtract the numbers in the same places.

Way to see and think

Even when the numbers get large, align the digits of the numbers according to their places.

Summary

For subtracting 3-digit numbers in vertical form, align the digits of the numbers according to their places and then subtract the numbers in the same places.

1 Let's calculate the following in vertical form.

① $768-534$ ② $879-412$ ③ $647-317$ ④ $965-864$

2 Let's think about how to calculate $342 - 128$ in vertical form.

Yu

I can't do $2 - 8$ for the ones place.

＼ Want to think ／

Purpose Can we think it as the same way as in the subtraction of 2-digit numbers in vertical form?

Sara

$$\begin{array}{r} 342 \\ -\ 128 \\ \hline \end{array} \rightarrow \begin{array}{r} {\scriptstyle 3\ 10} \\ 3\not{4}2 \\ -\ 128 \\ \hline 4 \end{array} \rightarrow \begin{array}{r} {\scriptstyle 3\ 10} \\ 3\not{4}2 \\ -\ 128 \\ \hline 14 \end{array} \rightarrow \begin{array}{r} {\scriptstyle 3\ 10} \\ 3\not{4}2 \\ -\ 128 \\ \hline 214 \end{array}$$

Way to see and think

Borrow 1 ten from the tens place as 10 ones.

1 Let's explain how to calculate $825 - 451$ in vertical form.

$$\begin{array}{r} 825 \\ -\ 451 \\ \hline \end{array} \rightarrow \begin{array}{r} 825 \\ -\ 451 \\ \hline 4 \end{array} \rightarrow \begin{array}{r} {\scriptstyle 7\ 10} \\ \not{8}25 \\ -\ 451 \\ \hline 74 \end{array} \rightarrow \begin{array}{r} {\scriptstyle 7\ 10} \\ \not{8}25 \\ -\ 451 \\ \hline 374 \end{array}$$

Summary Even in the case of borrowing in 3-digit numbers, we can borrow one from the higher place as we learned.

Akari

2 Let's calculate the following in vertical form.

① $363 - 114$ ② $432 - 217$ ③ $540 - 513$

④ $629 - 351$ ⑤ $257 - 183$ ⑥ $905 - 375$

? Is there a case when we have borrowing twice?

3 Let's think about how to calculate $425-286$ in vertical form.

\ Want to think /

Purpose What should I do if I can't do subtraction for the ones place nor the tens place?

Subtraction algorithm for $425-286$ in vertical form

Hundreds	Tens	Ones

$$\begin{array}{r} 425 \\ -\ 286 \\ \hline \end{array}$$

Borrow 1 ten from the tens place as 10 ones

① $15-6$

Borrow 1 hundred from the hundreds place as 10 tens.

② $11-8$

③ $3-2$

1 Let's calculate the following in vertical form.

① $424-185$ ② $821-373$ ③ $510-176$

④ $420-235$ ⑤ $242-64$ ⑥ $740-69$

4 Let's think about 305 − 178 in vertical form.

Hundreds	Tens	Ones

$$\begin{array}{r} 305 \\ -\ 178 \\ \hline \end{array}$$

Borrow 1 hundred from the hundreds place as 10 tens.

Borrow 1 ten from the tens place as 10 ones.

① 15 − 8

② 9 − 7

③ 2 − 1

$$\begin{array}{r} \overset{2}{\cancel{3}}\,\overset{\overset{9}{\cancel{10}}}{0}\,\overset{10}{5} \\ -\ 1\ 7\ 8 \\ \hline \end{array}$$

$$\begin{array}{r} \overset{2}{\cancel{3}}\,\overset{\overset{9}{\cancel{10}}}{0}\,\overset{10}{5} \\ -\ 1\ 7\ 8 \\ \hline 7 \end{array}$$

Summary When we can't subtract in the same place, we can borrow from the higher place.

Sara

1 Let's calculate the following in vertical form.

① 405 − 286 ② 402 − 107 ③ 800 − 197

5 Let's think about how to calculate 1000 − 895 in vertical form. ▷

	1	0	0	0
−		8	9	5

1 Let's calculate the following in vertical form.

① 1000 − 536 ② 1041 − 784 ③ 1237 − 414

? Can we add and subtract using larger numbers?

3 Calculation of larger numbers

Let's think about how to find the answer for larger numbers by using what you have already learned.

Can we calculate the thousands place?

Can we carry and borrow in the same way as we have learned?

? \ Want to think /

Even when we deal with larger numbers, can we add or subtract in vertical form as we have learned so far?

① 4175 + 3658

	4	1	7	5
+	3	6	5	8

② 6073 + 2981

	6	0	7	3
+	2	9	8	1

③ 7008 + 2992

	7	0	0	8
+	2	9	9	2

④ 3925 − 1947

	3	9	2	5
−	1	9	4	7

⑤ 3007 − 2639

	3	0	0	7
−	2	6	3	9

⑥ 10000 − 5089

	1	0	0	0	0
−		5	0	8	9

! **Summary** Way to see and think

Even when we deal with larger numbers, whenever we start adding and subtracting the numbers from the ones place, we can get the answer.

1 Let's calculate the following in vertical form.

① 4563 + 3125 ② 2606 + 3198 ③ 3587 + 6413

④ 6497 − 2135 ⑤ 8114 − 3518 ⑥ 10000 − 6001

4 Thinking of ways to calculate easily

Let's think about how to calculate $298+120$.

We can use vertical form.

Yu

Can't we calculate in easier ways?

Sara

? Purpose \ Want to improve /

Can we add or subtract in easier ways?

① Let's explain Sara's idea on the right.

Sara's idea

I thought $298+120$ as the same as $300+118$.

Haruto

Why is she making 120 to 118?

1 Let's calculate $500-198$ in easier ways.

! Summary

In addition, the answer does not change by adding a number to the augend and subtracting that same number from the addend.

$298+120=$ ☐

+2 ↓ ↓ −2

$300+118=$ ☐

In subtraction, the answer does not change by adding the same number to both the subtrahend and the minuend.

$500-198=$ ☐

$502-200=$ ☐

2 Let's calculate the following in easier ways.

① $308+197$ ② $499+350$ ③ $199+299$

④ $301-99$ ⑤ $600-297$ ⑥ $200-95$

2 Let's think about how to calculate $875 + 47 + 53$.

\ Want to improve /

Purpose When adding 3 numbers, how can we calculate easily?

① Let's explain Akari's and Haruto's ideas.

Akari's idea

$875 + 47 = 922$. Then I added 53 to 922.

Haruto's idea

$47 + 53 = 100$.
So I calculated this first, and then did $875 + 100$.

Summary 【Rule of Associativity】

When adding 3 numbers, changing the order of addition gives the same answer.

$$875 + 47 + 53 = 875 + (47 + 53)$$

If you change the order of calculations, it can be easier.

1 Let's calculate the following in easier ways.

① $492 + 84 + 16$　　② $52 + 365 + 48$

2 Let's calculate the following mentally. How can we calculate easily? Let's explain Yu, Sara, and Akari's ideas.

① $35 + 46$　　② $81 - 27$

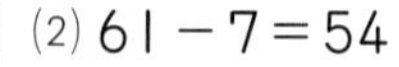

In ①, if we think 46 as 50, it means that I have added 4.

3 Let's calculate the following mentally.

① $18 + 6$　　② $68 + 29$　　③ $23 - 8$　　④ $71 - 46$

CAN What can you do?

We can add in vertical form. → pp.58 ~ 61, p.68

1 Let's calculate the following in vertical form.

① 324 + 253　② 146 + 537　③ 473 + 261
④ 246 + 485　⑤ 354 + 249　⑥ 464 + 368
⑦ 734 + 862　⑧ 947 + 587　⑨ 457 + 546
⑩ 4137 + 1425　⑪ 2056 + 3794　⑫ 2361 + 7639

☐ We can subtract in vertical form. → pp.64 ~ 68

2 Let's calculate the following in vertical form.

① 658 − 325　② 374 − 138　③ 546 − 369
④ 432 − 136　⑤ 604 − 247　⑥ 700 − 463
⑦ 1529 − 716　⑧ 1153 − 645　⑨ 1000 − 437
⑩ 3947 − 1925　⑪ 3142 − 1734　⑫ 10000 − 4005

☐ We can calculate in easier ways. → pp.69 ~ 70

3 Let's calculate the following in easier ways.

① 397 + 240　② 800 − 198
③ 5387 + 57 + 43　④ 26 + 3285 + 74

☐ We can make a math expression and find the answer. → p.68

4 Ayumi has 3596 yen and her sister has 4487 yen in their savings.

① Who has more saving and by how much?
② How much are their total savings?

Supplementary Problems → p.133

Which "Way to See and Think Monsters" did you find in "4 Addition and Subtraction"?

When I was dealing with addition and subtraction of larger numbers, I found "Same Way."

Yu

When I was seeking for easier ways to calculate...

Haruto

Utilize Usefulness and Efficiency of Learning

1 Let's fill in each □ with a number.

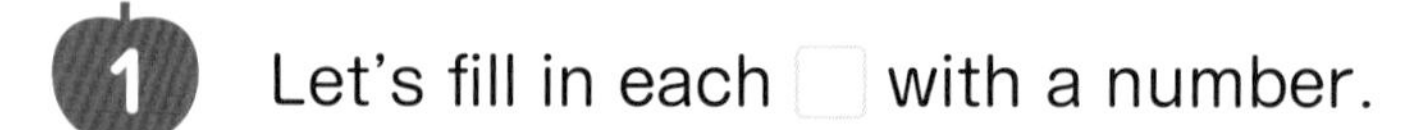

①
```
   1 □ 5
 + □ 7 □
 -------
   6 4 3
```

②
```
   3 □ 9
 + □ 3 □
 -------
   6 0 0
```

③
```
   5 □ □
 + □ 9 6
 -------
   7 0 4
```

④
```
   8 5 □
 - 2 □ 5
 -------
   □ 8 5
```

⑤
```
   □ □ 3
 -   4 □
 -------
     6 6
```

⑥
```
   9 □ □
 - □ 5 3
 -------
     8 7
```

2 Let's find the mistakes in the following calculations and write the correct answers.

①
```
   2 8 7
 + 1 4 9
 -------
   4 2 6
```

②
```
   4 0 3
 - 2 4 6
 -------
   1 6 7
```

3 There are two 1000-yen bills, three 100-yen coins, four 10-yen coins and four 1-yen coins in the wallet. Let's think about the following.

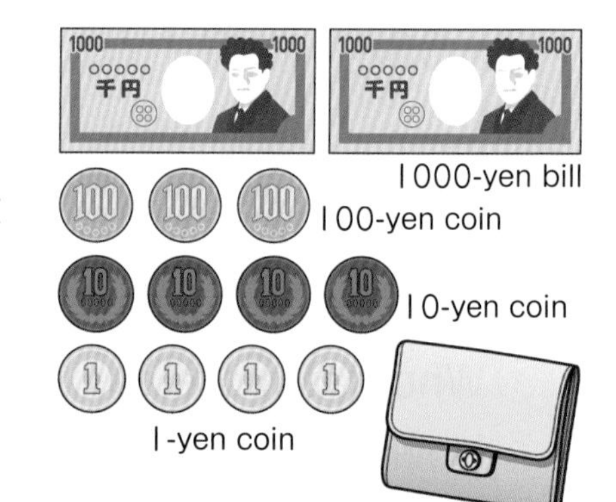

① I will use 733 yen for shopping. Which coins should I use to pay in order to get the change only using 100-yen coins?

② I will use 733 yen for shopping. Which coins should I use to pay in order to get the change only using 500-yen coins?

③ I will use 538 yen for shopping. Which coins should I use to pay in order not to get the change of 1-yen coins?

With the Way to See and Think Monsters...

Let's Reflect!

Let's reflect on which monster you used while learning "4 Addition and Subtraction."

Same Way

In additions and subtractions with large numbers, we could calculate in the same way by thinking per place.

① What are the similarities among addition and subtraction of 2-digit numbers and those of 3-digit numbers?

【Addition of 2-digit numbers】

$$\begin{array}{r} 38 \\ +\ 27 \\ \hline 65 \end{array}$$

【Addition of 3-digit numbers】

$$\begin{array}{r} 248 \\ +\ 187 \\ \hline 435 \end{array}$$

【Subtraction of 2-digit numbers】

$$\begin{array}{r} 45 \\ -\ 27 \\ \hline 18 \end{array}$$

【Subtraction of 3-digit numbers】

$$\begin{array}{r} 425 \\ -\ 286 \\ \hline 139 \end{array}$$

We could calculate addition and subtraction of 2-digit and 3-digit numbers in the same way as we learned by aligning the digits of the numbers according to their places.

Sara

We could use the same idea for carrying and borrowing too.

Yu

Let's deepen. → p.142

? Solve the ?

Even when the numbers get larger, we could calculate addition and subtraction in vertical form in the same way as we learned.

Sara

Want to Connect

Are there vertical forms for multiplication and division?

Akari

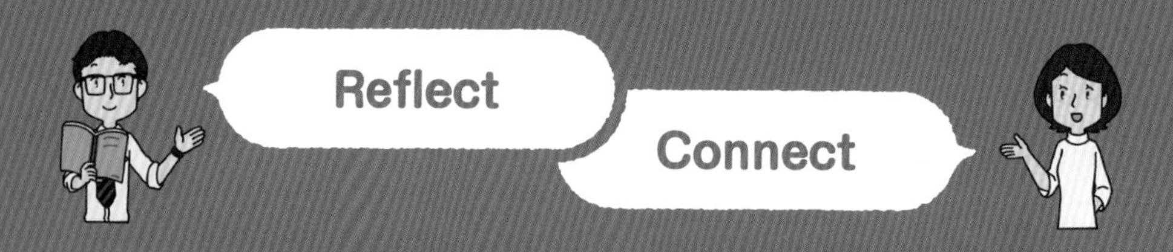

Problem Let's calculate in vertical form.

① 499 + 498

$$\begin{array}{r} 499 \\ +\ 498 \\ \hline 997 \end{array}$$

3 smaller than 1000.

add 3

502 + 498 = 1000

as a result

add 3

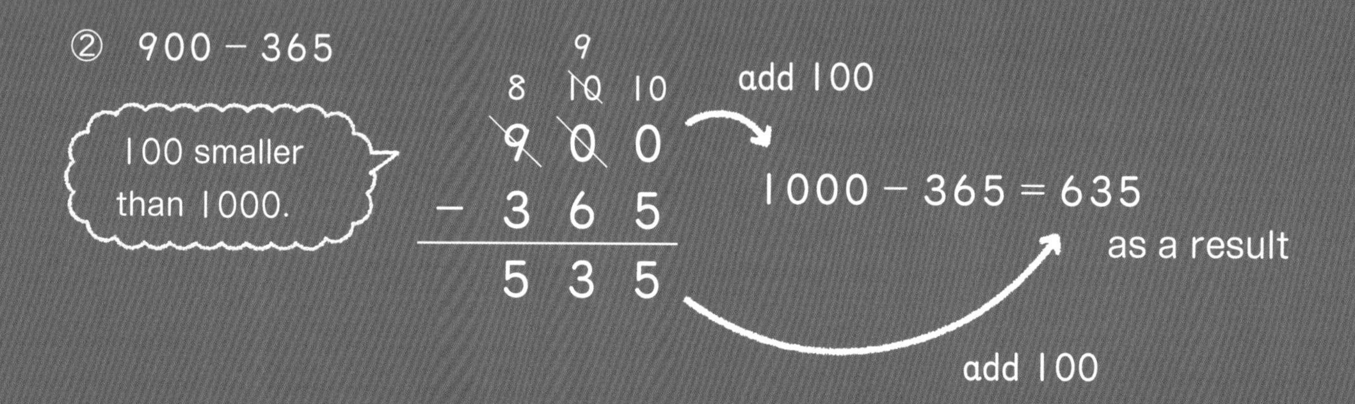

② 900 − 365

$$\begin{array}{r} 900 \\ -\ 365 \\ \hline 535 \end{array}$$

100 smaller than 1000.

add 100

1000 − 365 = 635

as a result

add 100

Can I calculate 4-digit numbers in vertical form?

Haruto

Remember what you have learned to calculate in vertical form.
Let's try to do in the same way.

① Add the numbers in each place
② Starting from the ones place
③ (Addition) Carry 1 to the next place
(Subtraction) Borrow 1 from a higher place

Let's confirm in the ways we have learned.

```
    5 0 2
+   4 9 8
---------
  1 0 0 0
```

I could do it in the same way!

```
    9  9
   1̸0 1̸0 10
  1̸ 0  0  0
-    3  6  5
------------
     6  3  5
```

I could do it in the same way!

Summary

- We can do addition that has an answer larger than 1000 and subtraction from 1000 in the same way as we have learned.
- I think we can calculate numbers larger than 1000 in the same way.

Sara
We can calculate by aligning the number according to the places.
We can calculate 4-digit numbers in the same way.

Want to Connect

I want to try calculating various other numbers.
Yu

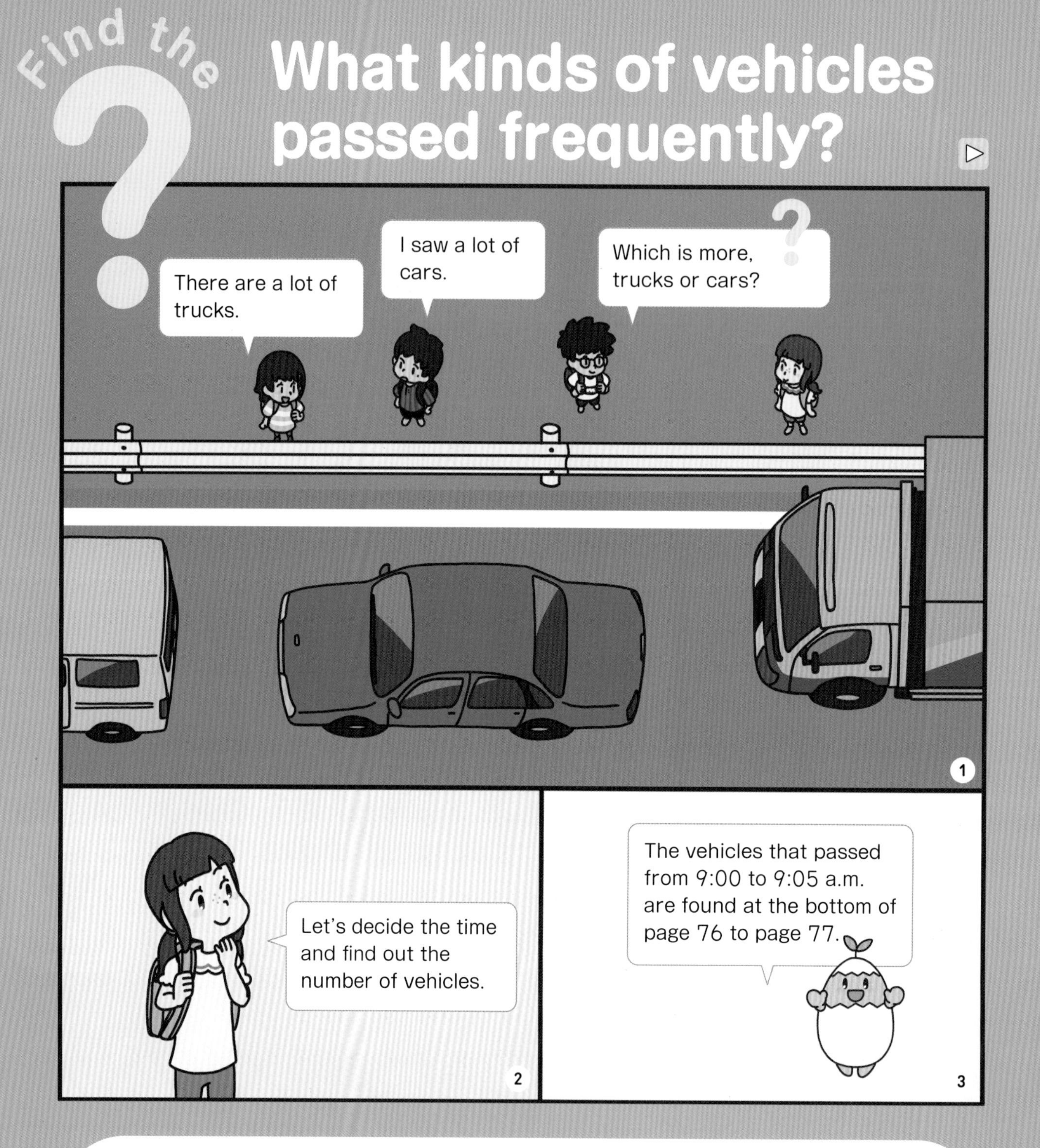
Find the
What kinds of vehicles passed frequently?
There are a lot of trucks.
I saw a lot of cars.
Which is more, trucks or cars?
1
Let's decide the time and find out the number of vehicles.
2
The vehicles that passed from 9:00 to 9:05 a.m. are found at the bottom of page 76 to page 77.
3

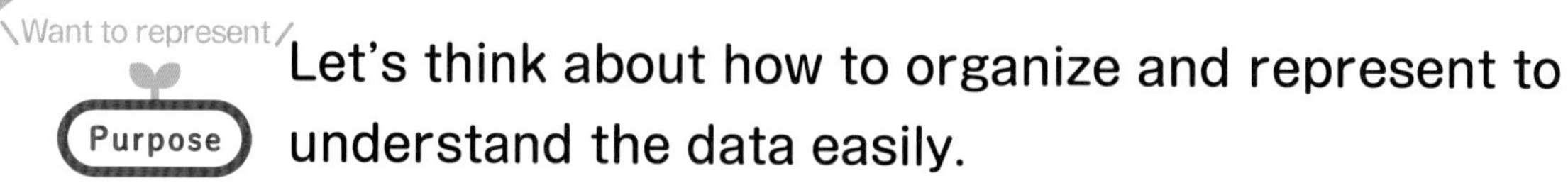
Want to represent
Purpose
Let's think about how to organize and represent to understand the data easily.

5 Tables and Graphs

Let's summarize investigations for easy understanding.

1 Tables

1 The table of the right shows the data about the vehicles that passed in front of our school from 9:00 to 9:05 a.m.

Vehicles that passed (In front of the school, 9:00 to 9:05 a.m.)

Kind	Number of Vehicles	
Truck	正	
Car	正 T	
Bus	T	
Ambulance	一	
Total		

Sara

We counted all of them in 2nd grade.

It's difficult to count vehicles because they are moving.

Haruto

① Let's change the character "正" to numbers.

一…1　　T…2　　下…3　　正…4　　正…5

② Let's discuss how to make a table.

③ Which kind of vehicle passed the most frequently and by how many?

④ Let's write the total number of vehicles.

1 Let's investigate the vehicles that passed in front of the school from 9:00 to 9:10 a.m. on the same day as **1**. The table below shows the data in **1**.

The cars that passed between 9:05 and 9:10 on the same day as **1** are on pages 79 to 109.

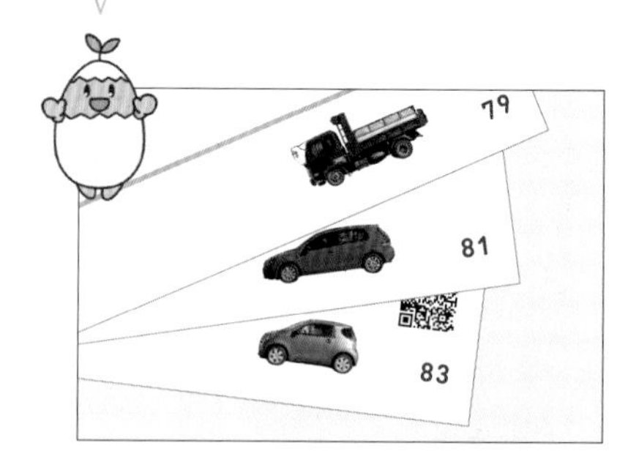

① Let's discuss what "Others" in the table on the right represents.

Akari

There are vehicles that are not seen frequently...

Vehicles that passed (in front of the school, 9:00 to 9:10 a.m.)

Kind	Number of vehicles	
Truck	正	
Car	正 丅	
Bus	丅	
Others	一	
Total		

② Let's look at the pictures found at bottom of pages 79 to 109 to know the vehicles that passed from 9:05 to 9:10 a.m. Add the data obtained by using the table on the right.

③ How many vehicles passed from 9:00 to 9:10 a.m? Where should the number should be written in the table?

④ What can we learn from this table?

The car most frequently seen in front of the school is...

Yu

Vehicles like an ambulance that passed infrequently are included in "others."

Summary

It is easier to count and organize in a table by representing the numbers in "正".

Way to see and think

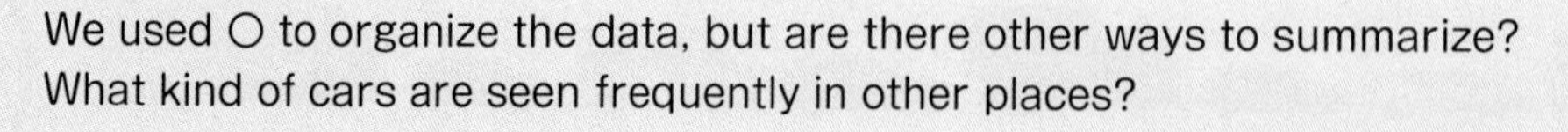

We used ○ to organize the data, but are there other ways to summarize?
What kind of cars are seen frequently in other places?

2 Bar graph

1 The bar graph on the right shows the results of the investigation on the previous page. Let's explore the graph.

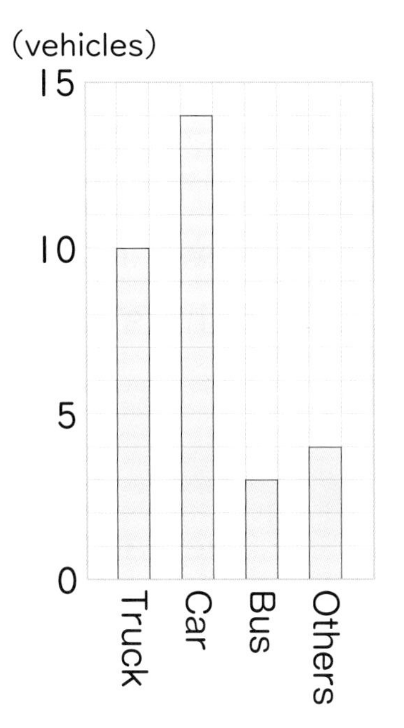

We used ◯ in 2nd grade to represent numbers on the bar graph. — Yu

Want to explore

Purpose What can we find out from the bar graph?

1. What does the length of bars represent?
2. How many vehicles are represented by 1 cell on the graph?
3. How many trucks were there?
4. What kind of vehicles passed most frequently? How many were there?

A graph, which uses bars of different lengths to represent data, is called a **bar graph**.

Title → Vehicles that passed (in front of the school, 9:00 to 9:10 a.m.)

Unit → (vehicles)

Vertical Axis

Horizontal Axis

15, 10, 5, 0 — Truck, Cars, Bus, Others

Sara: In a bar graph, we can compare the size of numbers easily.

Haruto: It looks different from a table.

⑤ The bar graph in the previous page was changed into the one on the right. What has changed in the graph?

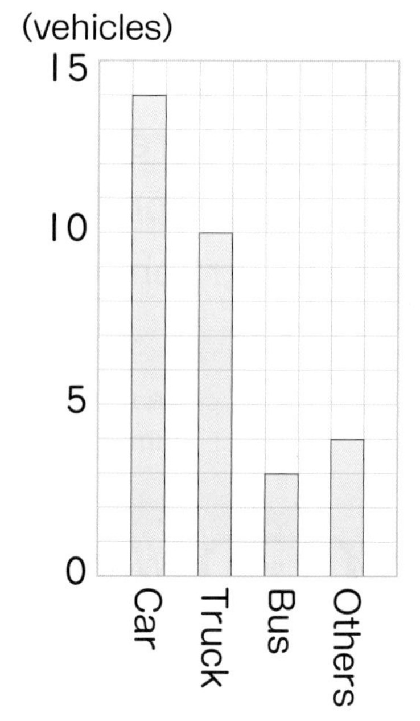

In the bar graph, the bars are often drawn in descending order. The "Others" bar is usually drawn last.

Investigation of vehicles in front of the station →

2

The bar graph on the right shows the results of the investigation about the number of the vehicles that passed in front of the station. Let's explore the graph.

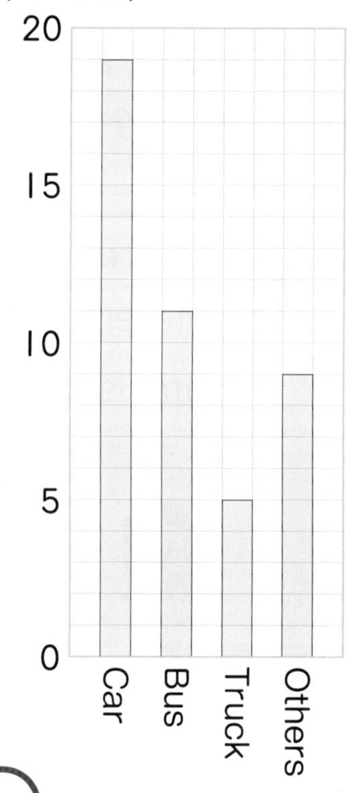

① What is the difference in number between the most and the least? Don't include the "Others."

② What is the total number of vehicles that passed in front of the station?

③ What can you say about this graph compared to the graph above which investigated the vehicles that passed in front of the school?

④ What other things would you like to investigate based on what you have found so far?

Summary

It is easier to understand what is most and what is least by using a bar graph.

3

Sayuri and her friends investigated the number of children who visited the nurse's office because of illness or injury in April, and recorded the number of children who visited each weekday and presented the data in a bar graph.

① In what order does the graph represent the data?

\ Want to know /

Purpose Why is it not drawn in descending order?

Sara

Children who visited the nurse's office in a week

(children)

0 5 10 15

Mon
Tue
Wed
Thu
Fri

② How many children are represented by 1 cell on the graph?

③ Let's find out the number of children who visited the nurse's office each day.

④ Let's compare the numbers of children who visited the nurse's office on Monday and Thursday.

A bar graph is sometimes drawn horizontally.

When the data to be presented in bar graph are in order, like the days of the week, the bars should be drawn in that order.

1 In the graphs below, let's find out the number that represents 1 cell.

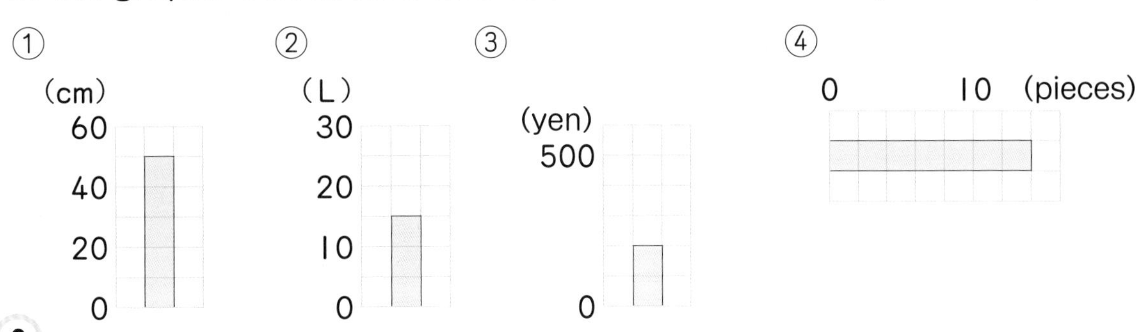

? Can we draw a bar graph by ourselves?

4

The table on the right shows the favorite sports of the 3rd grade children in class 1. Let's draw a bar graph.

Favorite sports

Sport	Number of children
Soccer	14
Baseball	10
Dance	7
Swimming	3
Others	2
Total	36

\ Want to represent /

Purpose How can we draw a bar graph?

How to draw a bar graph ▷

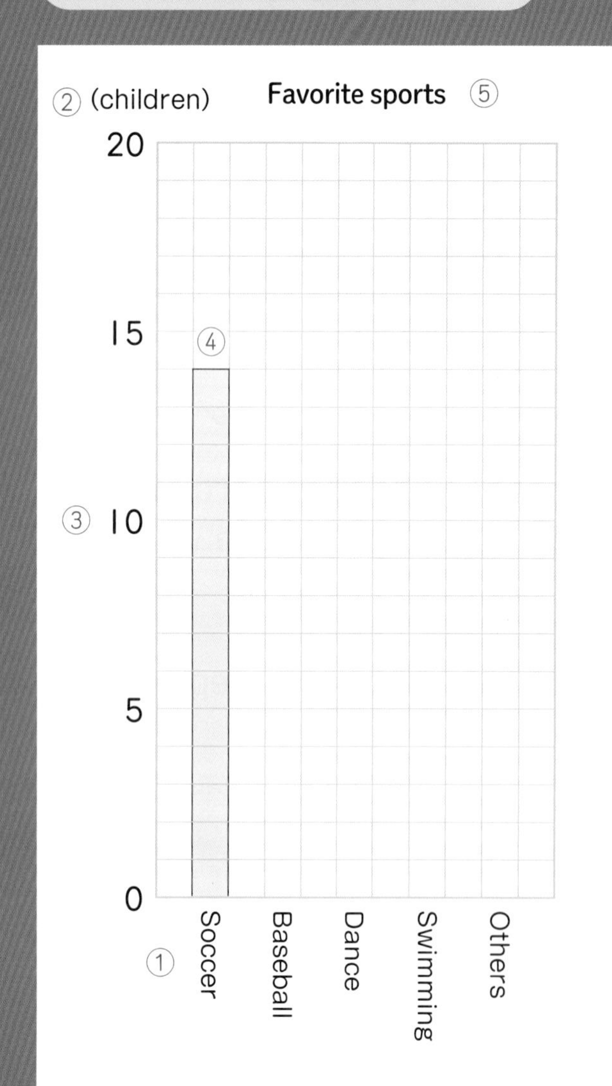

① Write each sport on the horizontal axis.

② Write the unit on the vertical axis.

③ Write the scale on the vertical axis. When deciding on the scale to use, consider the highest number of children, and write the units like 5, 10, etc.

④ Draw each bar according to the number of children.

⑤ Write the title of the graph.

1 The following table shows the number of the 3rd grade children in each class whose favorite sport is soccer. Let's draw a bar graph.

3rd grade children who like soccer

Class	Number of children
1	14
2	15
3	11
Total	40

We can decide whether to write the graph in the order of the class or write in descending order.

Yu

(children)
20
10

2 The following table shows the favorite sport of all the 3rd grade children. Let's draw a bar graph.

Favorite sports

Sport	Number of children
Soccer	40
Baseball	35
Dance	15
Swimming	10
Others	5
Total	105

Way to see and think

How many children shall we set for 1 scale on the graph?

? How can we draw a graph that is easy to read?

I heard that there will be new books in the library.

If we can know what kind of books are popular, we can decide what books to stock.

3 Putting tables together

1 The following tables show the number of the 3rd grade children who borrowed each kind of books in April, May, and June.

Books borrowed in April

Kind	Number of books
Fiction	15
Biography	6
Picture	8
Others	5
Total	

Books borrowed in May

Kind	Number of books
Fiction	21
Biography	19
Picture	24
Others	8
Total	

Books borrowed in June

Kind	Number of books
Fiction	16
Biography	14
Picture	19
Others	9
Total	

① What is the total number of books that were borrowed in each month?

② Which kind of books was borrowed in the most in each month?

In the tables above, it's not easy to find out the popular books...

Akari

? Purpose \ Want to explore / What kind of table is easy to understand the popular books?

③ Let's organize the tables for each month into one table.

Books borrowed by the 3rd grade children (books)

Kind \ Month	April	May	June	Total
Fiction	15	21	16	52
Biography	6	19		ⓓ
Picture	8			ⓔ
Others	5			ⓕ
Total	ⓐ	ⓑ	ⓒ	ⓖ

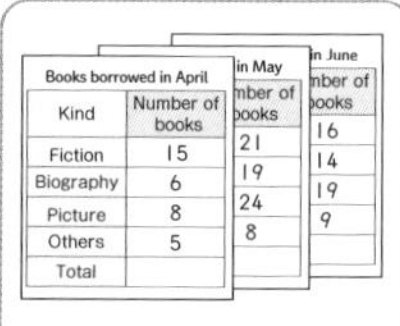

Books borrowed in April

Kind	Number of books
Fiction	15
Biography	6
Picture	8
Others	5
Total	

All we did was to put the 3 tables together.

Yu

④ How many fiction books were borrowed from April to June?

⑤ What numbers are in boxed ⓐ, ⓑ, ⓒ, ⓓ, ⓔ, and ⓕ?

⑥ What does the number in ⓖ represent?

⑦ What kind of book was borrowed the most from April to June?

What kind of books should I include in the library?

Sara

Summary

By putting some tables into one, it makes it easier to know what is the most and what is the least.

Way to see and think

1 The table below shows the number of children in Yuma's school who got injured and the type of injury they had in April, May, and June. Let's answer the following.

① How many children were injured in each month?

② What type of injury happened the most from April to June?

③ What does the number 46 in the table represent?

Record of injuries of children (children)

Kind \ Month	April	May	June	Total
Scratch	29	27	13	
Bruise	21	46	30	
Cut	13	7	4	
Sprain	7	4	2	
Others	10	14	6	
Total				

CAN What can you do?

□ We can read the numbers from tables or graphs and draw graphs. → pp.77 ~ 83

1 Children collected empty cans at Koharu's school. The following table and bar graph show the number of cans collected by the children in each grade.

Empty cans collected by children

Grade	1	2	3	4	5	6	Total
Number of cans		120		240	160		

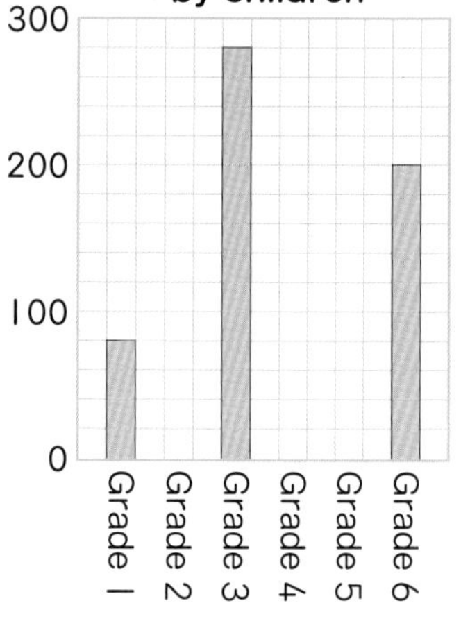

① How many cans does 1 cell represent on the graph?
② Let's fill in the blanks of the table above.
③ Let's draw the bars for the 2nd, 4th, and 5th grades on the graph.
④ Between the table and the graph, which makes it easier to see children in which grade collected empty cans the most?

□ We can read the numbers from tables. → pp.84 ~ 85

2 The table on the right shows the number of children in each grade at Ren's school who got injured and the kind of injury they had in June.
Let's answer the following.

Record of injuries of children (June) (children)

Kind \ Grade	1	2	3	4	5	6	Total
Scratch	3	ⓑ	2	5	3	4	21
Cut	ⓐ	2	2	3	ⓔ	3	ⓖ
Bruise	1	1	ⓒ	2	2	ⓕ	13
Others	2	3	1	1	0	2	9
Total	7	10	8	ⓓ	9	13	ⓗ

① Write the numbers that apply in boxes ⓐ, ⓑ, ⓒ, ⓓ, ⓔ, ⓕ, ⓖ, and ⓗ.
② What kind of injury happened the most in June?
③ In which grade did the children get injured the least in June?

Supplementary Problems → p.135

Which "Way to See and Think Monsters" did you find in "5 Tables and Graphs"?

I found "Summarize" when I was trying to organize the findings of the investigation.

Akari

When I was thinking about the amount of 1 cell in the graph...

Yu

With the Way to See and Think Monsters...

Let's Reflect!

Let's reflect on which monster you used while learning "5 Tables and Graphs."

Summarize

By summarizing the findings into tables or graphs, we could see the data easily.

① What kind of table or graph would be appropriate to summarize the followings?

I want to find out the most popular fruit in our class.
Sara

I want to find out what kind of books are borrowed among each grade in our school.
Haruto

Unit

By changing the amont of 1 cell, we could draw graphs to make the data easier to understand.

② Let's think about what you notice regarding the following two graphs.

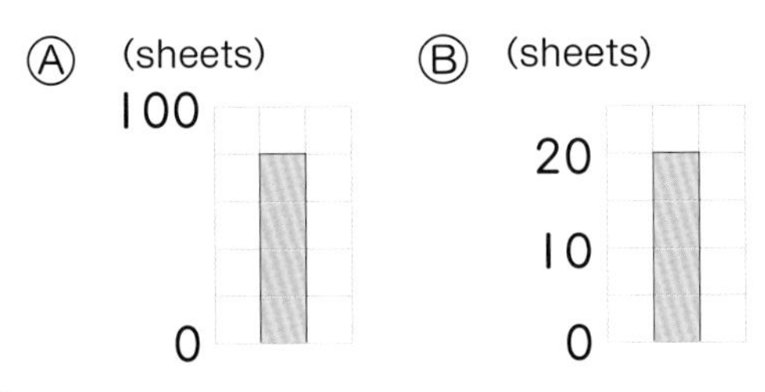

The length of the graph is the same, so does it mean both of them represent the same amount?
Akari

1 cell for graph Ⓐ represents ☐ sheets. 1 cell for graph Ⓑ represents ☐ sheets. So, Ⓐ is more.
Yu

Let's deepen. → p.144

Solve the ?

By putting the tables together and by using bar graphs, it was easier to organize the data of the investigation.

Want to Connect

Are there other ways to organize the data other than the ways we have learned so far?

Find the ?

How far did the rubber band car run?

Let's think about how to measure a length between 2 places longer than 1 m.

Length

Let's think about the units of longer length and how to represent it.

1 How to measure

1 Let's think about how to measure the length that a rubber band car ran.

Yu: I think we can measure it by using a 1 m ruler several times.

Akari: Can we measure in a straight line?

It is difficult to measure by using 1 m rulers.

The length between two places along a straight line is called **distance**.

Tape measures are good tools for measuring the distance.

Be careful when you adjust a 0 point of the measure to one end.

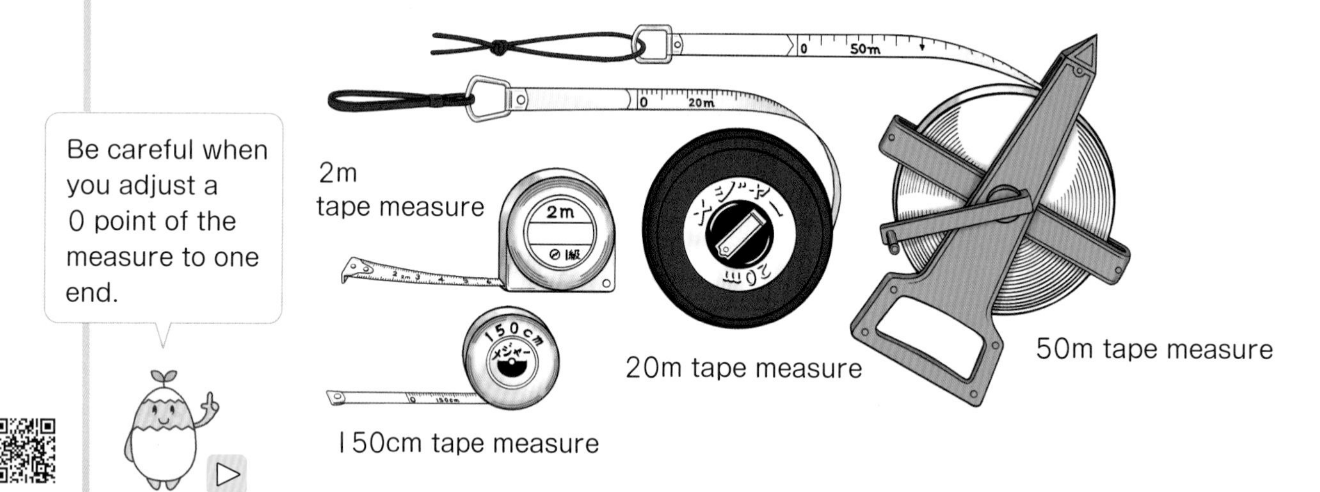

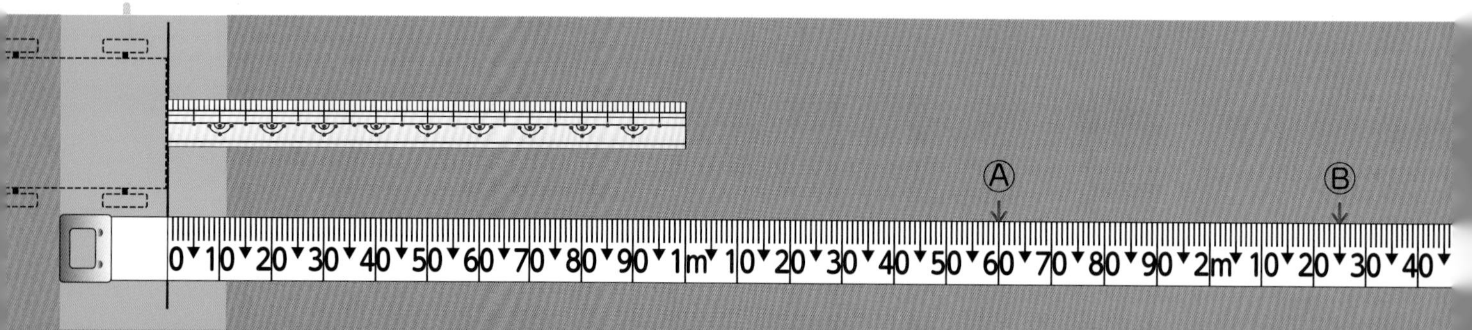

1 Let's investigate how to use a tape measure.

Some tape measures have a 0 point, others don't.

① Look at the location of 0 m on the tape.

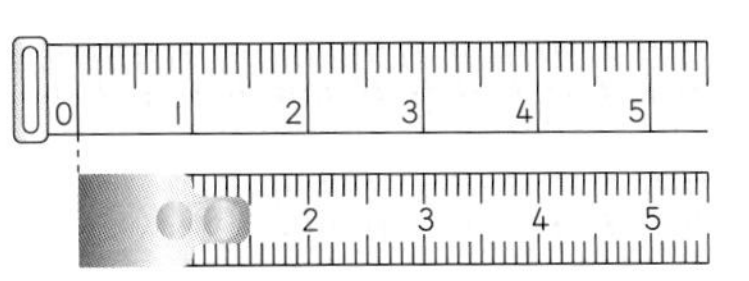

② Let's find the distance that Yuki's car ran by reading her record shown in the tape measure below.

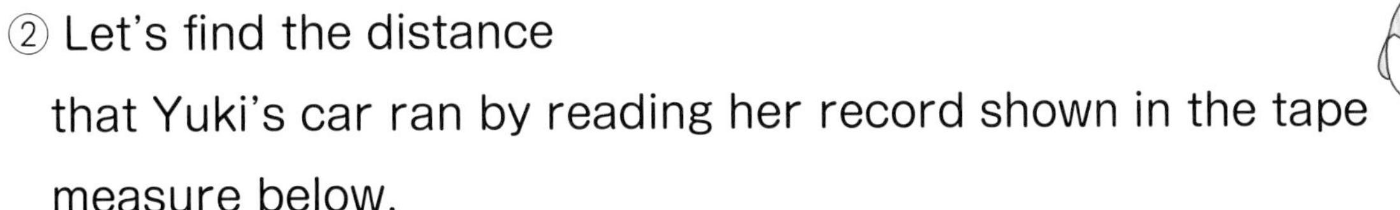

③ What is the length in m and cm are the lengths of Ⓐ, Ⓑ, and Ⓒ shown on the tape measure below? Let's point ↓ at each scale that represents the length Ⓓ, Ⓔ, and Ⓕ.

Ⓓ 3m5cm Ⓔ 75cm Ⓕ 4m54cm

Summary

When we measure a distance longer than 1m, we can use a tape measure.

2 About how long is 10m?

Walk to a point that you think is 10m away.

Then let's measure the actual length.

? Can we measure the length of various objects using a tape measure?

Yuki's record

0 70 80 90 3m 10 20 30 40 50 60 70 80 90 4m 10 20 30 40 50 60 70 80 90 5m

\ Want to try /

Purpose First, estimate each length.

2 For the rulers and tape measures in the ☐ below, which is appropriate to measure the following length?

Ⓐ the length and width of a desk　Ⓑ the length around a can

Ⓒ the length of a hallway

30cm ruler	1m ruler(meterstick)	150cm tape measure	50m tape measure

1 What is the length in cm around the tree in the picture on the right?

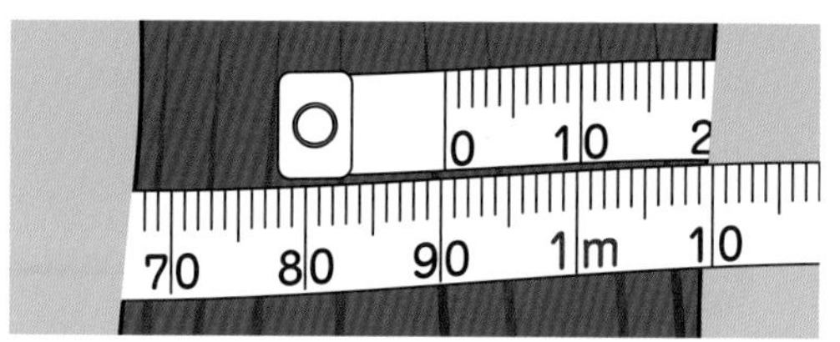

2 Let's estimate the lengths of objects that can be found in our surroundings, and actually measure them.

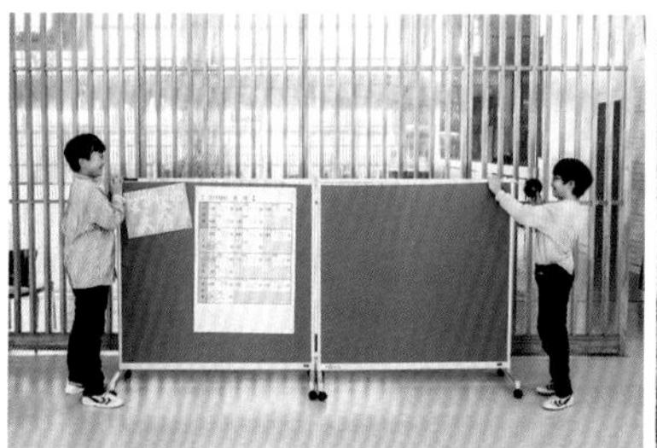

Which tree has the longest circumference in our school?

Haruto

Lengths of various objects

Objects measured	Estimated length	Actual length
Width of a bulletin board		
Height of an iron bar		
Length around a tree		

Write down what you found out and understood after the investigation.

? Can we represent longer length?

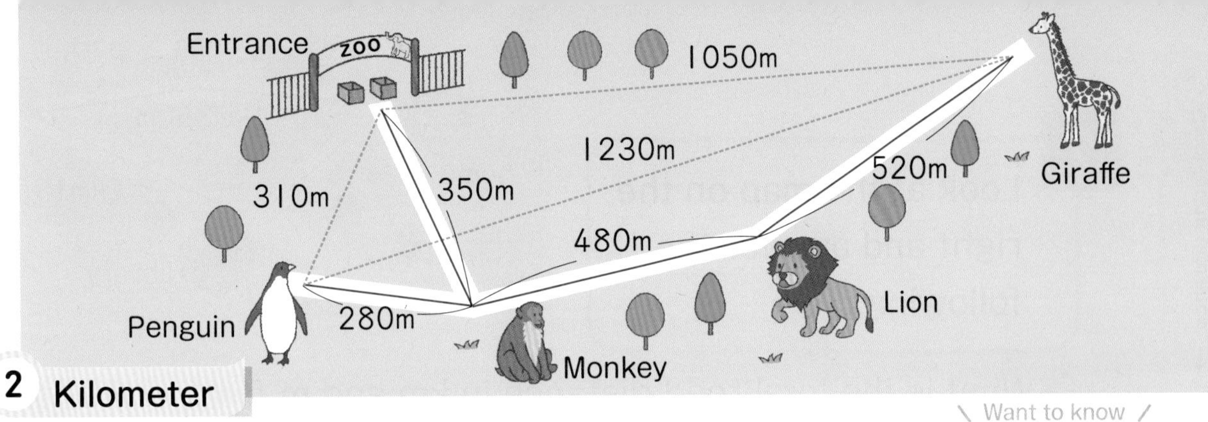

2 Kilometer

Kilometer →

1 Look at the map of the zoo above and answer the following.

\ Want to know /

Purpose What is the difference between distance and road distance?

Akari

① What is the distance in m from the entrance to the penguins? What is the distance in m measured along the road?

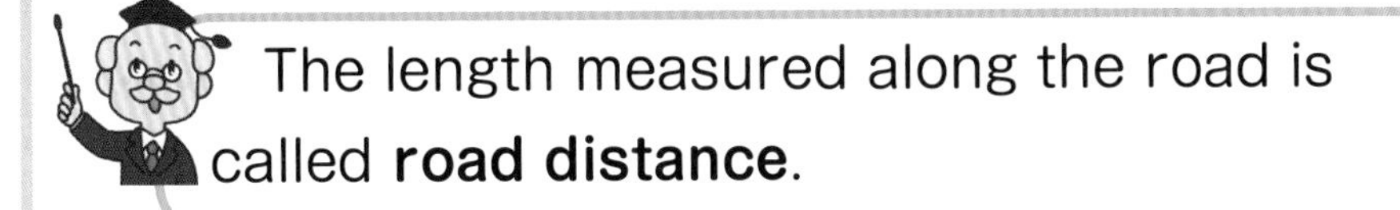

The length measured along the road is called **road distance**.

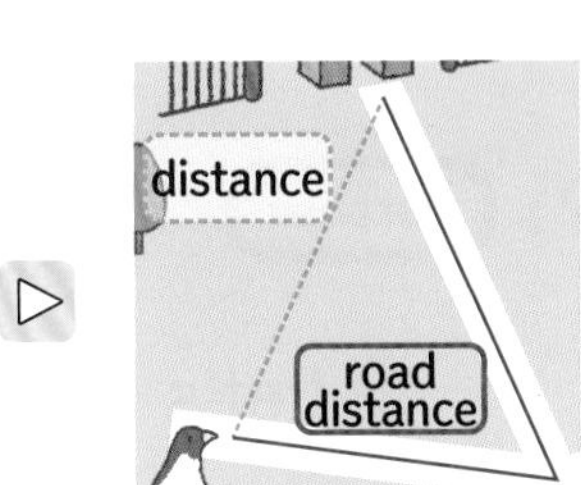

② What is the road distance in m from the monkeys to the giraffes?

1000m is written as 1km and is called one **kilometer**.

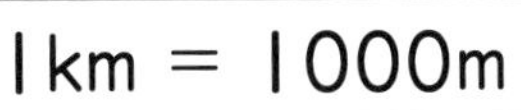

1km = 1000m

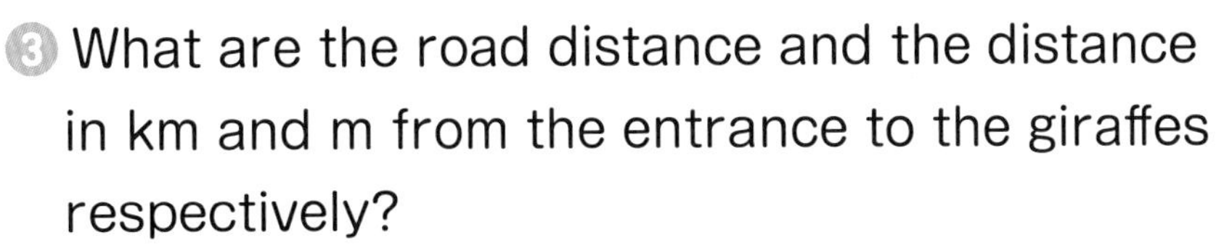

Way to see and think

③ What are the road distance and the distance in km and m from the entrance to the giraffes respectively?

km			m
1	3	5	0
1	0	5	0

road distance 1350m = ☐ km ☐ m

distance 1050m = ☐ km ☐ m

1km350m is said as "1 kilometer and 350 meters."

④ What are the road distance and the distance in km and m from the penguins to the giraffes respectively?

? Can we calculate km as we have learned?

2 Look at the map on the right and answer the following.

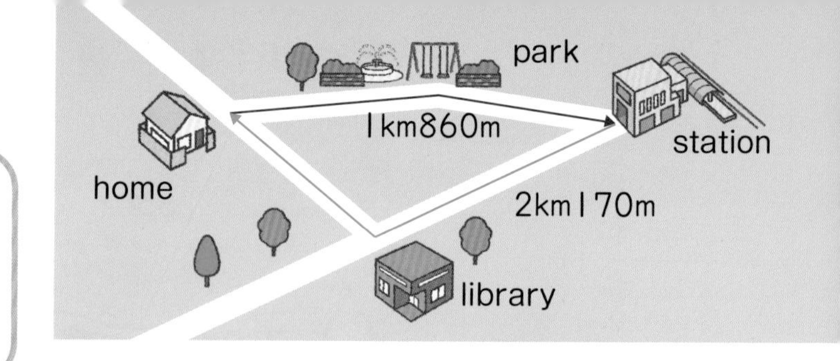

① What is the total road distance in km and m from home to the station through the front of the park and returning home through the front of the library?
Let's write a math expression.

\ Want to think /

? Purpose How can we calculate distance longer than 1km?

② Let's explain the ideas of Sara's and Yu's.

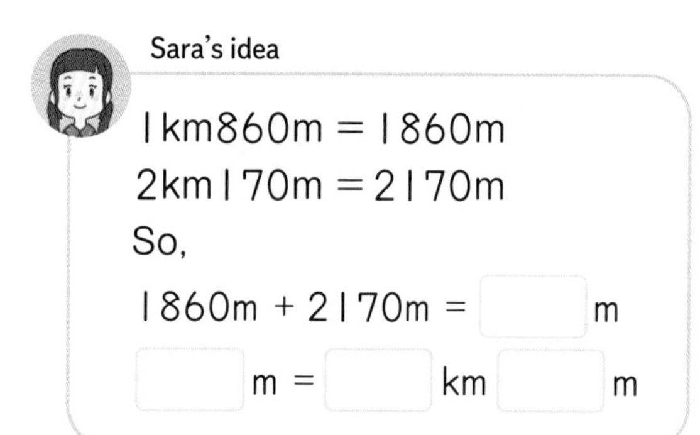

Sara's idea
1km860m = 1860m
2km170m = 2170m
So,
1860m + 2170m = m
m = km m

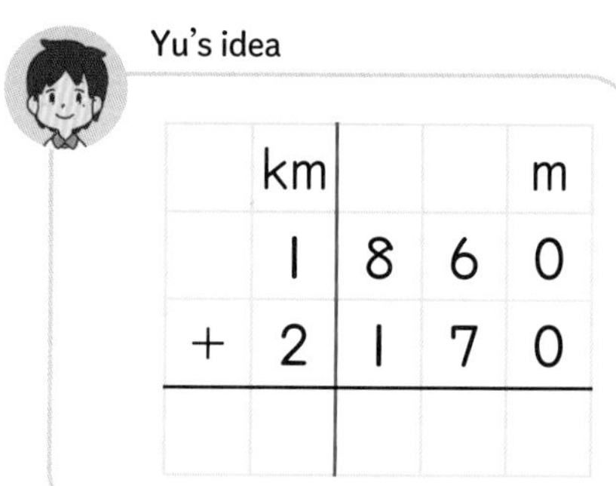
Yu's idea

	km			m
	1	8	6	0
+	2	1	7	0

Way to see and think
Can we calculate as we did in cm and m?

③ Which road distance from the home to the station is longer, and by how much?

	km			m
	2	1	7	0
−	1	8	6	0

! Summary

We can calculate longer length by aligning the units and by adding or subtracting numbers of the same units.

Way to see and think

3 Let's actually walk 1km around the school.

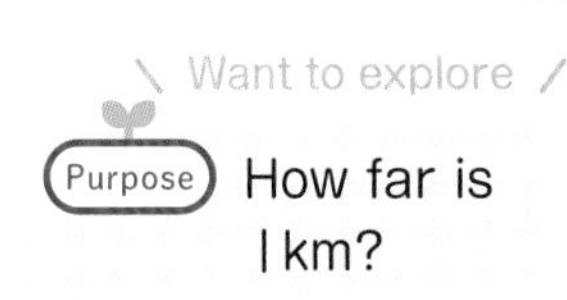

Want to explore
Purpose How far is 1km?

Yu

① Where do you think you will end up if you walk 1km from school?

② Let's estimate how long it takes you by walking 1km.

③ Let's summarize your observations.

Let's walk 1km	July 6
① Up to where did we walk?	
	Post office
② Time spent: 20minutes	
③ My reflection:	
1km was longer than I had expected.	

How many meters is 1km?

Akari

Haruto

How many minutes will it take to walk 1km?

▶1 Let's look for the units of length that are used in our surroundings.

Evacuation Site
昭和記念公園
Showa Kinen Park
500m

Tachikawa City, Tokyo Metropolis

(Itsukushima Shrine, Miyajima)

(Atomic Bomb Dome)

(Hiroshima Station)

Takuma takes a tour of Hiroshima City by a tram. He leaves Hiroshima Station, visits both the Atomic Bomb Dome and the Hiroshima Port, and finally arrives at the Hiroden-miyajima-guchi.

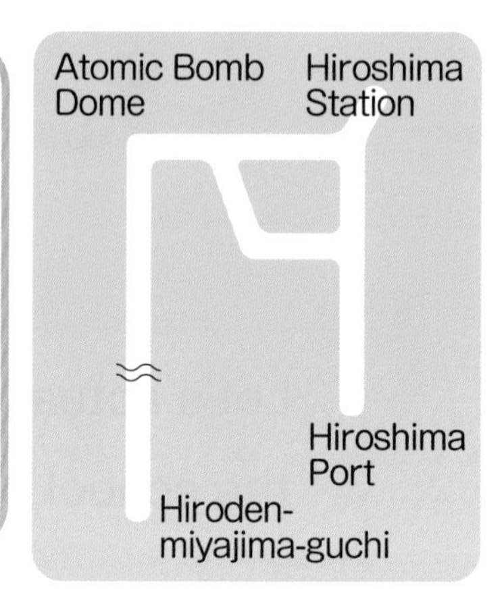

\ Want to think /

Purpose Which route should I take to travel all the places?

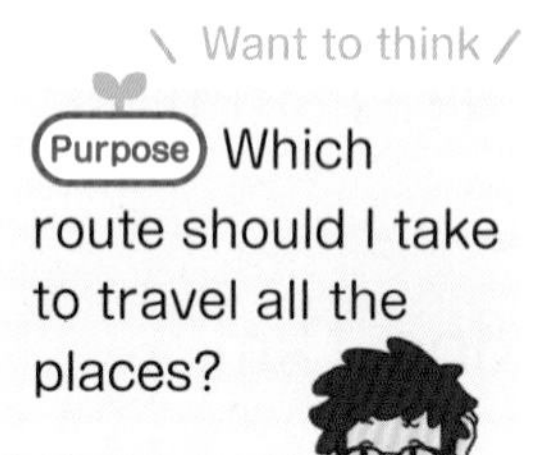
Haruto

Road distance and time

	Road distance	Time
Hiroshima Station ↔ Atomic Bomb Dome	2km400m	17min
Hiroshima Station ↔ Hiroshima Port	6km100m	32min
Atomic Bomb Dome ↔ Hiroshima Port	6km200m	36min
Atomic Bomb Dome ↔ Hiroden-miyajima-guchi	19km100m	51min

1. The table above shows the road distance and travel time between two places. Where should he go first in shorter time, to the Atomic Bomb Dome or to the Hiroshima Port?
2. Which will be a longer road distance, going first to the Atomic Bomb Dome or to the Hiroshima Port? How much longer is it?
3. In 2, which takes longer by tram and by how much?

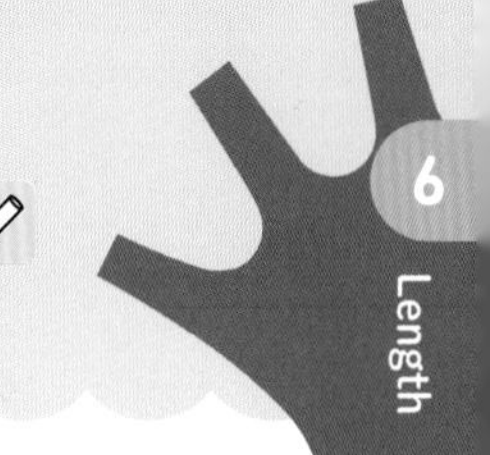

CAN What can you do?

□ We understand longer length. → p.90, p.93

1 Let's fill in the □ with numbers or words.

① Choose two places and measure the length between the two places along a straight line. This is called □.

② The distance measured along the road is the □.

□ We can read length correctly using a tape measure. → p.91

2 What is the length in m and cm of points Ⓐ, Ⓑ, and Ⓒ located on the tape measure below?

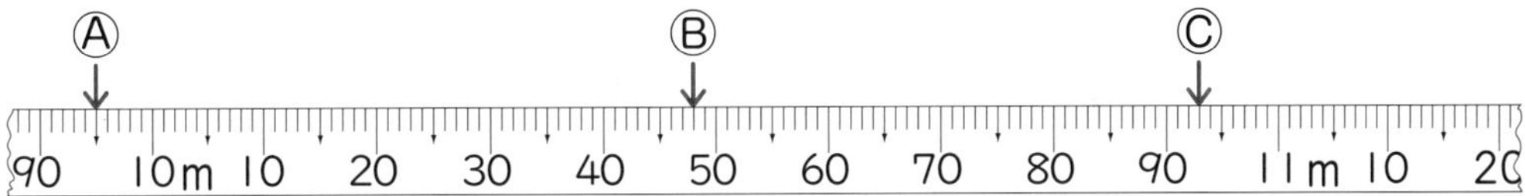

□ We understand the relationship between km and m. → p.93

3 Let's fill in each □ with a number.

① 1km = □ m ② 2km50m = □ m

③ 1116m = □ km □ m

□ We understand the difference between road distance and distance. → p.93

4 Look at the map on the right and answer the following.

① What is the road distance in km and m from Yuka's house to the school through the front of the park?

② What is the difference between the road distance in ① and the distance from Yuka's house to the school in m?

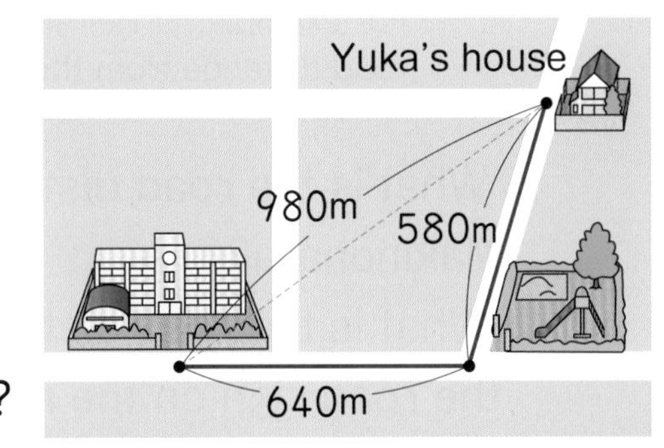

Supplementary Problems → p.137

Which "Way to See and Think Monsters" did you find in "6 Length"?

I found "Unit" when I was thinking about the unit of length.

Sara

Usefulness and Efficiency of Learning

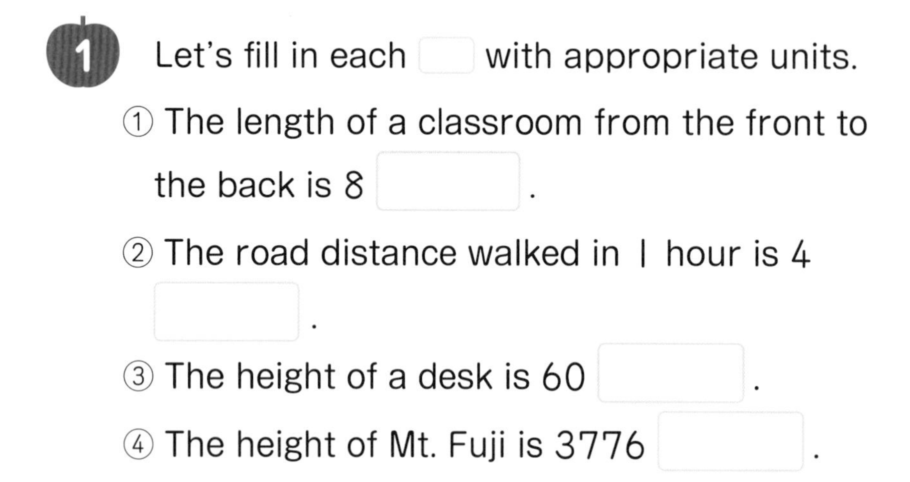

1 Let's fill in each ☐ with appropriate units.

① The length of a classroom from the front to the back is 8 ☐ .

② The road distance walked in 1 hour is 4 ☐ .

③ The height of a desk is 60 ☐ .

④ The height of Mt. Fuji is 3776 ☐ .

2 The road sign on the left was seen on Tomei Expressway Downlink from Tokyo to Nagoya. Then a road sign on the right appeared. Let's think about the following questions.

For example, "11 km" in the road sign on the left represents the road distance from the place where the road sign is to Yoshida.

① What is the road distance in km from Yoshida to Sagara-Makinohara?

② What is the road distance in km from the road sign on the left to the road sign on the right?

③ The diagram below shows the relationship of each places. Let's fill in the ☐ with symbols.

Ⓐ Yoshida Ⓑ Sagara-Makinohara Ⓒ Nagoya Ⓓ Kakegawa Ⓔ Fukuroi

☐ ☐ ☐ ☐ ☐ Tokyo

With the Way to See and Think Monsters...

Let's Reflect!

Let's reflect on which monster you used while learning in "6 Length."

Unit

By setting 1 km, 1 m, 1 cm, and 1 mm as one unit, we could represent various length in numbers.

① Let's summarize the relationship of units for representing length.

1000m = 1 [ ]
100cm = 1 [ ]
10mm = 1 [ ]

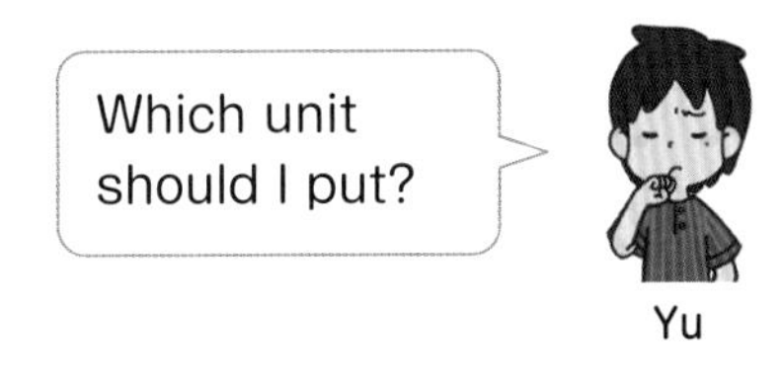

Align

We could calculate length by using the same units and aligning them when writing them down.

② Let's summarize how we calculated length.

1 km = 1000m, so I represented the following length using m only.

1 km 860 m = [ ]
2 km 170 m = [ ]
[ ] + [ ] = [ ]

I aligned the units, and calculated as follows.

	km			m
	1	8	6	0
+	2	1	7	0

Solve the ?

A tape measure was useful to measure long distance.

Want to Connect

It looks similar to the units such as L, dL, and mL. Are there any relationship between them?

Utilizing Math for SDGs

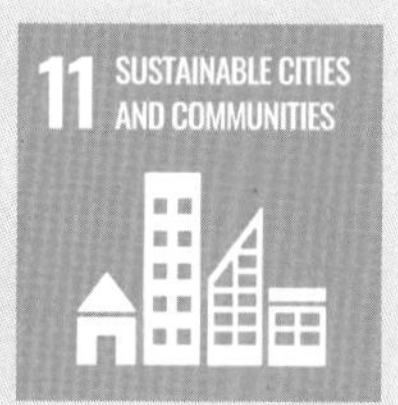

Watch out for traffic accidents!

You may be doing more activities with your friends, such as going to and from school, going out to play, and going to lessons. At such times, there is a risk of being involved in a traffic accident if you do not follow the traffic rules. Let's think about what we can do to avoid traffic accidents.

① The following table summarizes the number of traffic accidents per time zone. Let's summarize what you noticed.

Number of traffic accidents per time zone (number of accidents)

	6:00-8:00	8:00-10:00	10:00-12:00	12:00-14:00	14:00-16:00	16:00-18:00	18:00-20:00	20:00-22:00	22:00-0:00	0:00-2:00	2:00-4:00	4:00-6:00	Total
Grade1, 2	18	23	27	35	87	80	24	1	0	1	0	1	297
Grade3, 4	10	21	14	24	99	118	28	2	0	0	0	2	318
Grade5, 6	4	9	13	31	59	94	33	4	2	0	0	1	250
Total	32	53	54	90	245	292	85	7	2	1	0	4	865

② The table on the right summarizes the locations where traffic accidents occurred.Look at it and think about the places where you should watch out for traffic accidents on your way to school, around your house, or around your school.

Location of the traffic accident

Location	Number of accidents
Intersection	383
Near an intersection	98
Streets other than at or near intersections	366
At or near a railroad crossing or others	18
Total	865

③ The table on the right shows the cause of the traffic accident while walking. What should we pay attention to in order to avoid traffic accidents? Let's discuss.

Causes of the traffic accident while walking

Cause	Number of accidents
Ignored the traffic light	7
Crossed a street that was not a sidewalk	16
Crossed in front of or behind a stopped car	7
Crossed where you are not allowed to cross	3
Playing on the street	7
Jumped out of the way	63
Did not follow other traffic rules	11
Obeyed traffic rules but was involved in an accident	249
Total	363

Sara

We may need to check the traffic rules again.

Haruto

I want to be careful not to jump out of the way...

Think back on what you felt through this activity, and put a circle.

Let's reflect on yourself!

	Strongly agree	Agree	Don't agree
① I could find out some factors to avoid traffic accidents.			
② By observing the time zone, location, and the cause of the traffic accident, I could think about what to be careful about avoiding traffic accidents.			
③ I could find out the information from the table.			

	Strongly agree
④ I am proud of myself because I did my best.	

Let's praise yourself with some positive words for trying hard to learn!

Find the Places with an equal distance?

Students are lined up to play the ball-toss game.
It's too far from here to the basket. It's not fair.
1
How can we arrange ourselves so that nobody stands far away from the basket?
2
How about arranging ourselves in the shape of a square?
3
How about lining up in a circle?
How can we make all the distances the same?
4

\ Want to explore /

What would the arrangement look like when all the distances to the basket are the same?

Circles and Spheres

Let's explore the properties of round shapes and how to draw them.

1 Circles

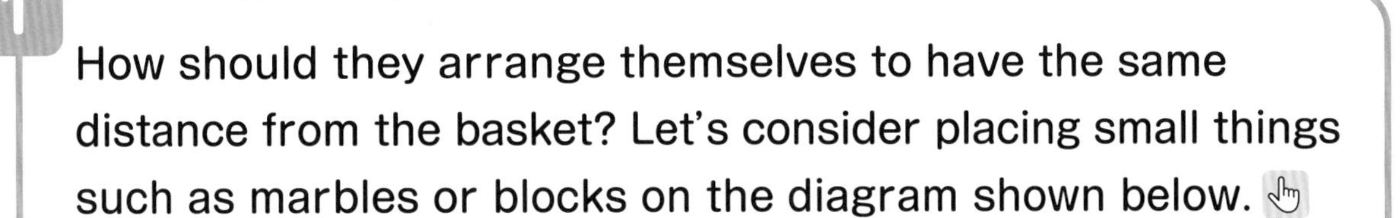

1
How should they arrange themselves to have the same distance from the basket? Let's consider placing small things such as marbles or blocks on the diagram shown below.

When they are arranged with the same distance from × at the center, the shape would be...

Yu

It would be something round.

Akari

1 Let's draw many points that are 3cm away from point A. What kind of shape will it become? ▷

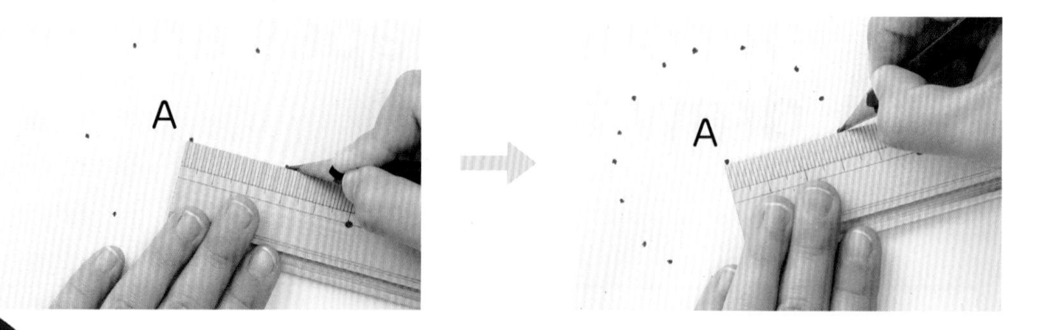

A round shape that consists of points that have the same distance from one point is called a **circle**. This point in the middle is called the **center** of the circle. The straight line from the center to any point of the circle is called the **radius**. In a circle, the length of every radius is the same.

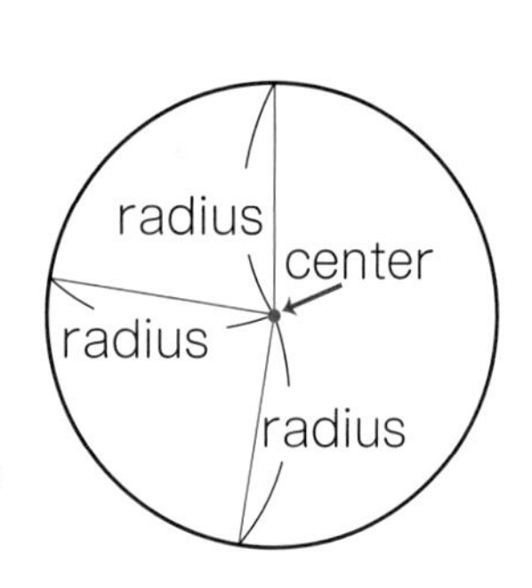

The circle that you have drawn in **1** has a radius of 3cm. Point A is the center of the circle.

Summary

Drawing many points at the same distance from a certain point becomes a round shape.

2 Let's draw many radii on the circle shown on the right and explore their length. Let's confirm that every radius has the same length.

3 Let's look for circular shapes in our surroundings.

(Fuji City, Shizuoka)

Way to see and think

Can we call any round shape such as ⬭ a circle?

? How can I draw a circle beautifully?

2 As shown on the right, let's draw circles of various sizes.

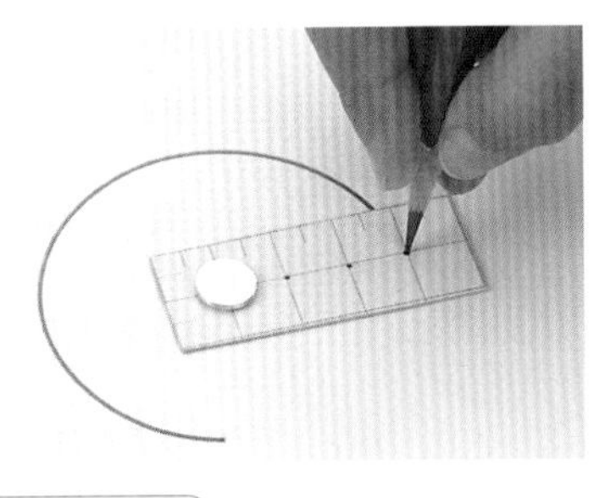

\ Want to try /

Purpose Can we draw circles of different size by changing the length of the radius?

Sara

Let's check whether the length of the radius are all the same.

1 Let's draw a circle with a radius of 2m by using a rope at our school ground.

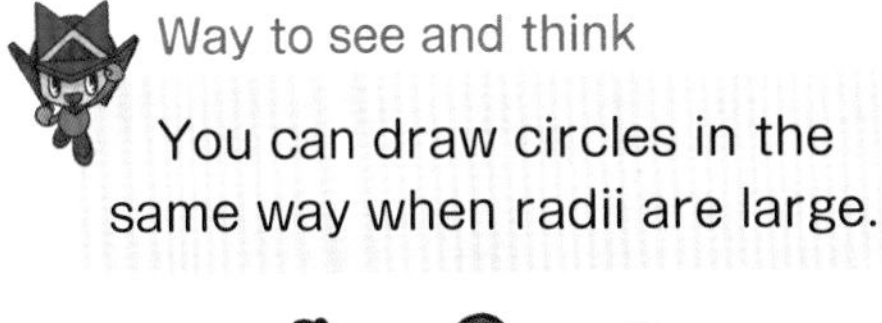
Way to see and think

You can draw circles in the same way when radii are large.

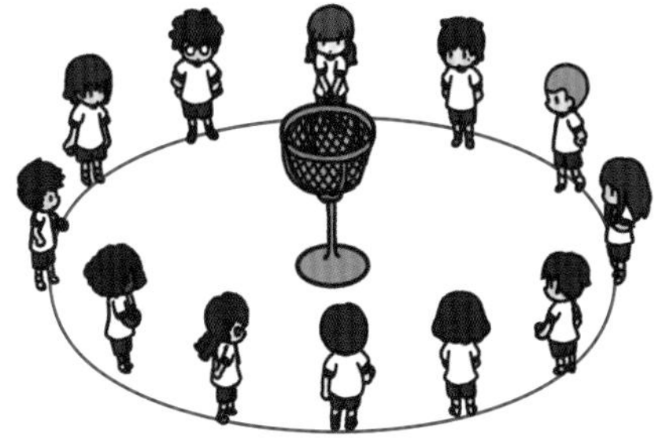

By drawing a circle, we can line up at the point where the length from the basket is all the same.

Akari

? Can we draw circles more easily?

3

It is useful to use a compass to draw circles. A circle with a radius of 1cm is drawn below. Using the same center, let's draw circles with a radius of 2cm, 3cm, 4cm, and 5cm.

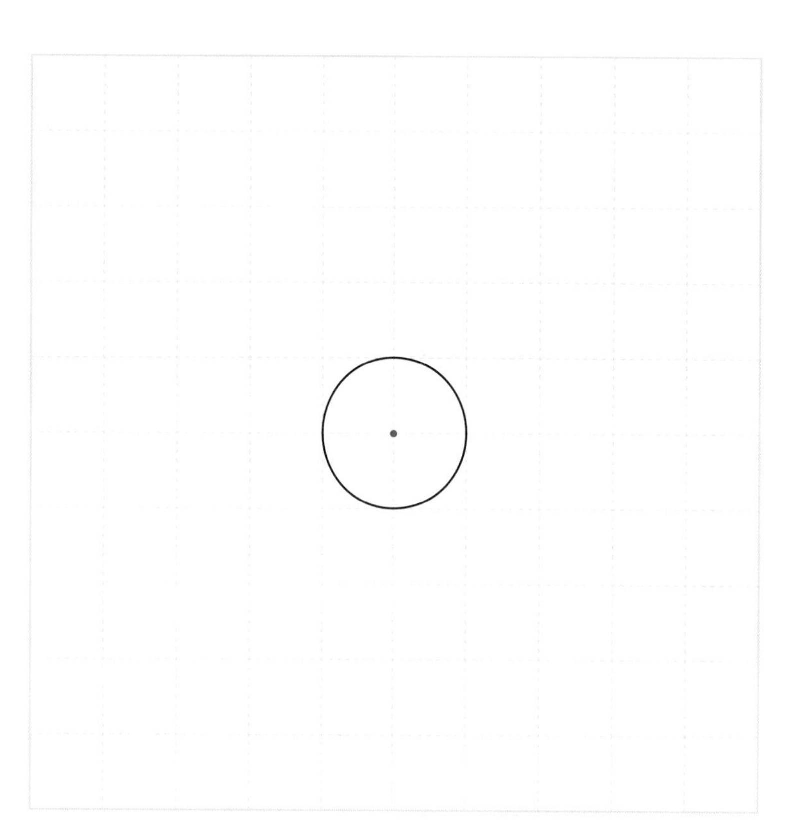

\ Want to know /

Purpose How can we use a compass to draw a circle?

Haruto

Let's draw circles using a compass referring to the instructions shown below.

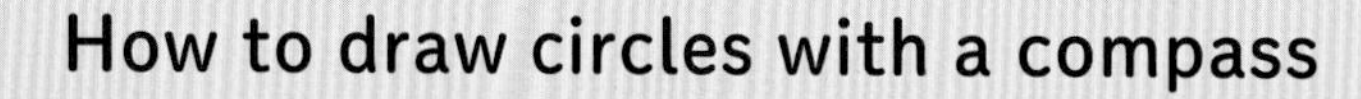
How to draw circles with a compass

(1) Remove any pad from underneath the page.

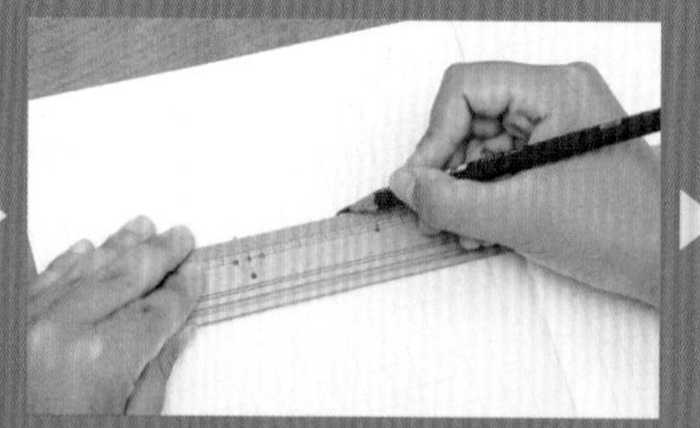

(2) First, determine the center of the circle, and then mark the points according to the length of the radius.

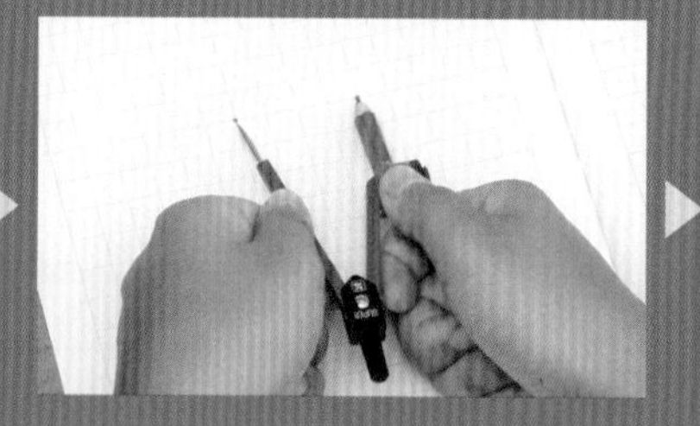

(3) Open the compass based on the length of the radius.

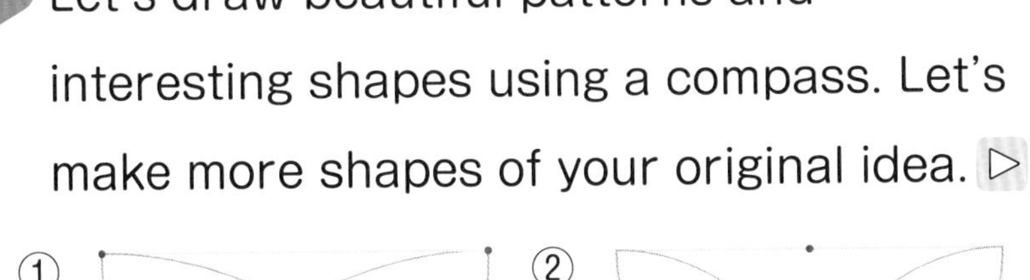

1 Let's draw beautiful patterns and interesting shapes using a compass. Let's make more shapes of your original idea. ▷

Tomihiro Art Museum (Midori City, Gunma Prefecture)

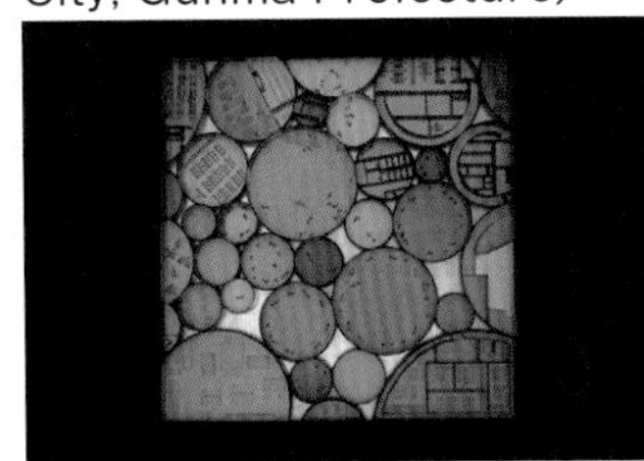

Model of Tomihiro Art Museum

① 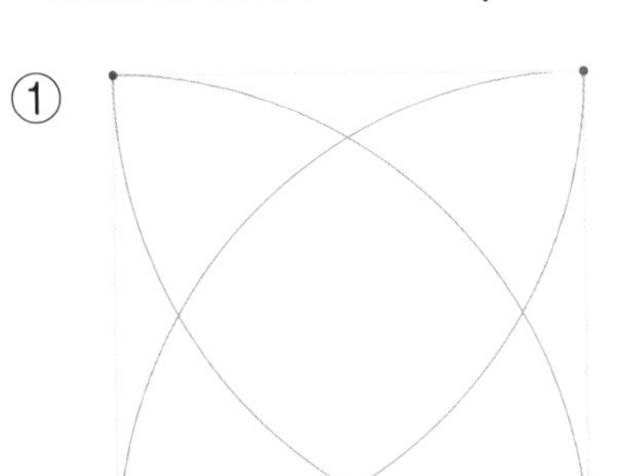

② 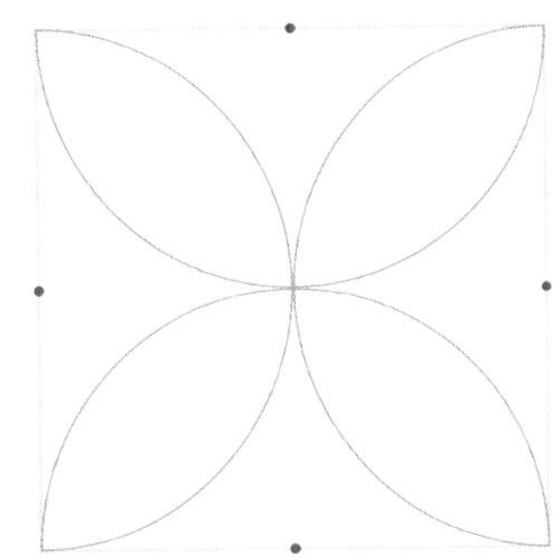

Yu: Where should I determine the center?

Akari: Let's find out where the radius is.

? Can you find the center of the circle?

It is good to turn your wrist toward yourself at first for a complete turn without stopping your compass.

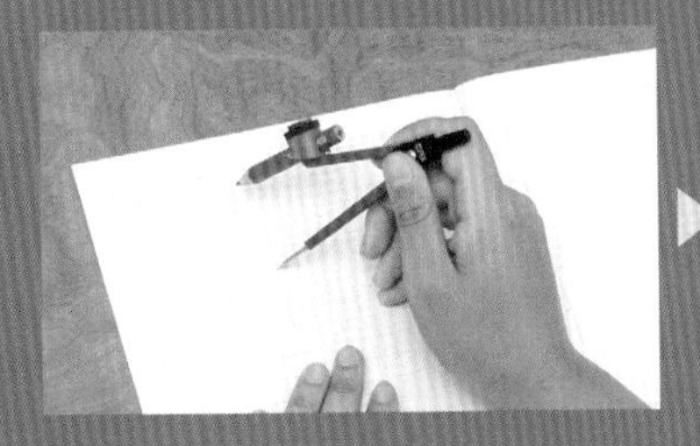

(4) Place the compass needle at the center.

(5) Start turning the compass toward your wrist.

(6) Turn the compass without stopping.

4

Let's draw a circle with the same size of the one shown on the right using a compass.

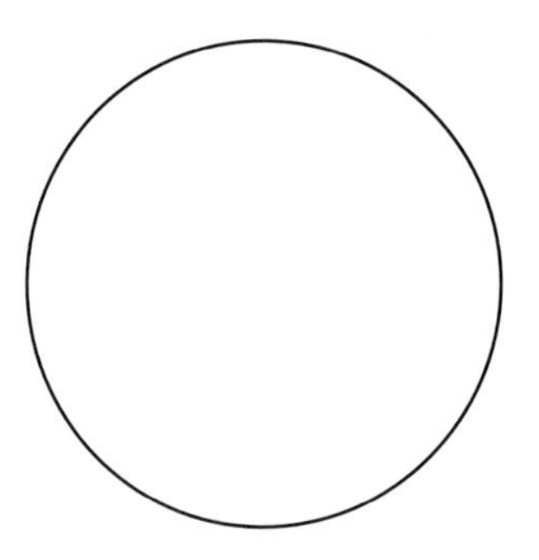

① What should I need to know to draw a circle?

How many cm is the radius?

Where should we place the needle of the compass?

\ Want to know /

? **Purpose** How can we find the center of a circle?

② Let's draw a circle, cut it out, and fold it as shown below. What can you understand from this?

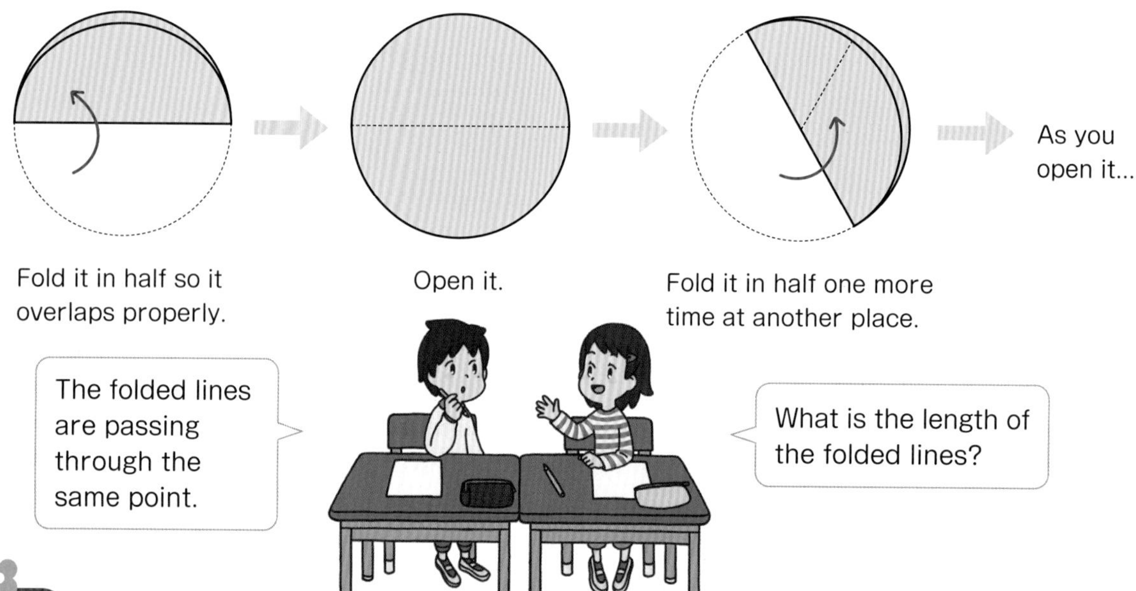

As you open it...
Fold it in half so it overlaps properly.
Open it.
Fold it in half one more time at another place.
The folded lines are passing through the same point.
What is the length of the folded lines?

! **Summary**

After folding two times a circle into halves that overlap properly, the center of the circle can be found at the point where the two folded lines intersect each other.

③ Let's find the center of the circle shown above and draw a circle with the same size.

1 Let's draw many straight lines from a point on the surrounding circle to other points on the surrounding circle as shown on the right. What kind of straight line is the longest straight line?

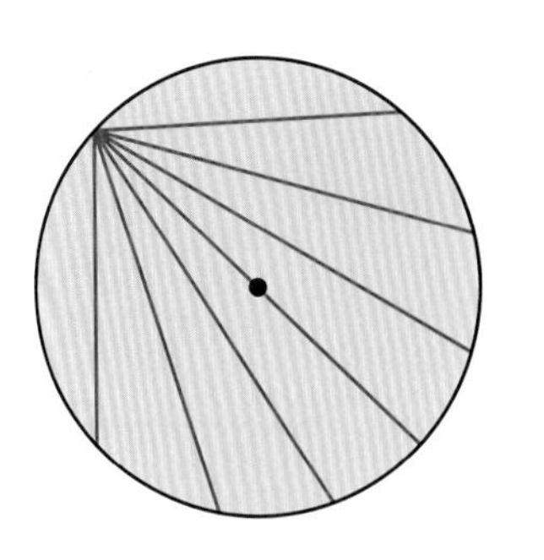

A straight line drawn from one point on the surrounding circle passing through the center of the circle to another point is called a **diameter** of the circle. The length of the diameter is twice the length of the radius.

There are many diameters that can be drawn in a circle and their lengths are all the same. All diameters pass through the center of the circle.

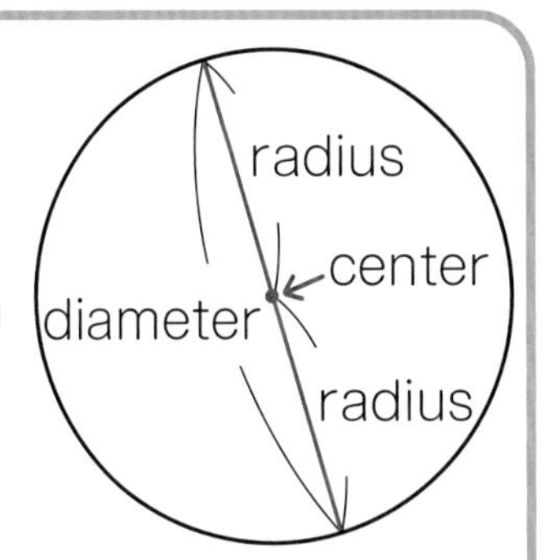

Diameter is the longest straight line among all the lines that could be drawn from the perimeter to perimeter of the circle.

Let's find ways to measure the diameter of a circle.

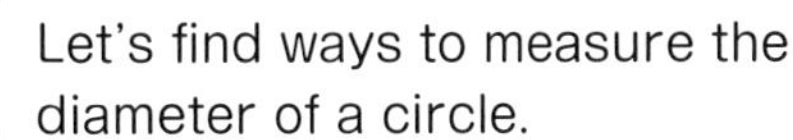

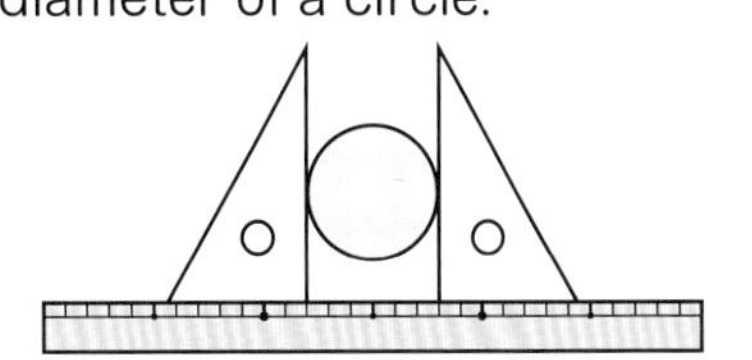

2 Let's draw a circle that fits exactly in a square with a side of 6cm.

① How many cm should the radius of the circle be?

② Let's find the center of the circle.

③ Let's draw the circle that fits exactly inside the square.

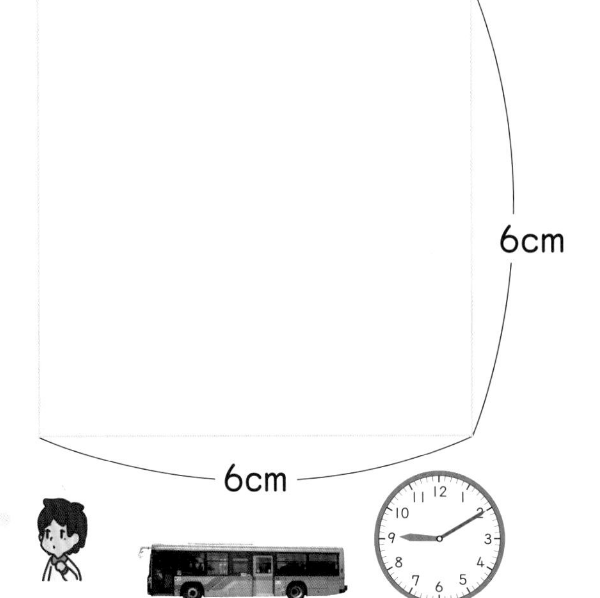

?

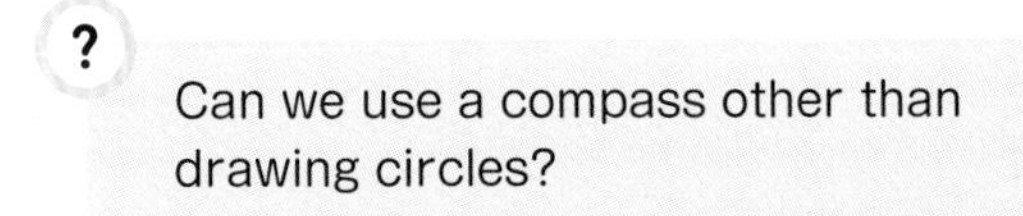

Can we use a compass other than drawing circles?

5

Takuto is going to the park as shown on the map below.
Which route is closer, Ⓐ or Ⓑ ?

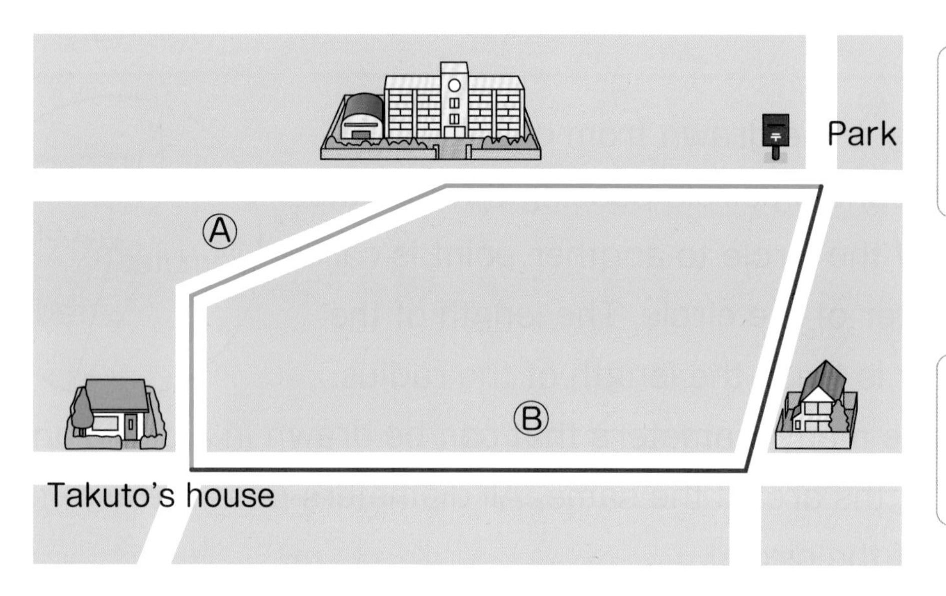

Can we compare them without using a ruler?

Yu

Can we compare lengths using a compass?

Sara

\ Want to compare /

Purpose How should we compare the length of bent lines?

❶ Let's transfer the length of the routes Ⓐ and Ⓑ to the following straight lines by using a compass, and explore which one is longer.

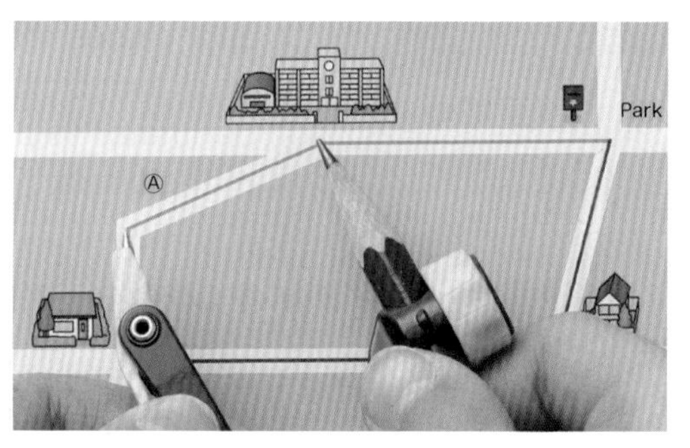

Ⓐ ____________________________

Ⓑ ____________________________

Summary

We can compare the length of routes by transferring each length to a straight line by using a compass.

A compass can be used to cut lines into the same length.

1 Let's cut the following straight line into segments of 3cm by using a compass.

2 Let's compare the length of Ⓐ, Ⓑ, and Ⓒ by using a compass.

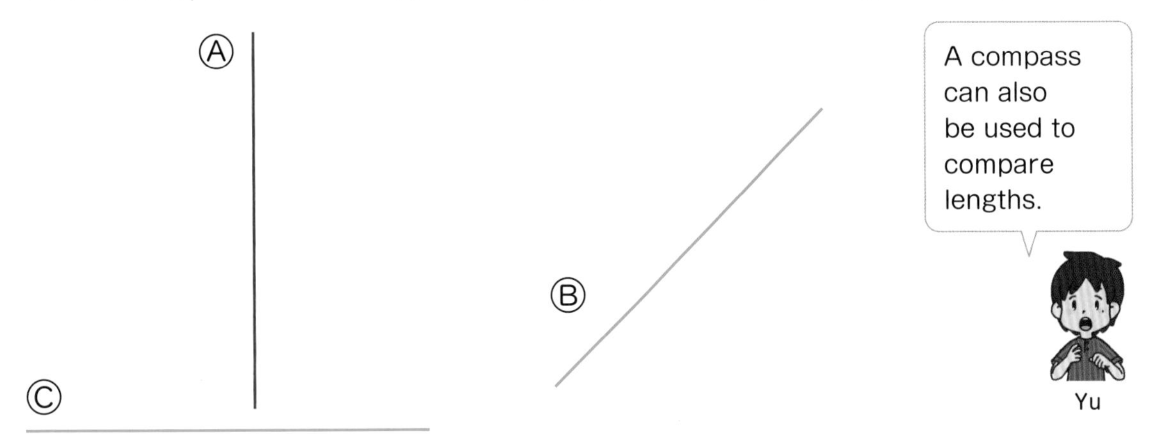

Ⓐ
Ⓑ
Ⓒ
A compass can also be used to compare lengths.
Yu

3 Let's use a compass and explore the followings in the diagram below.

How can I use the compass?
Haruto

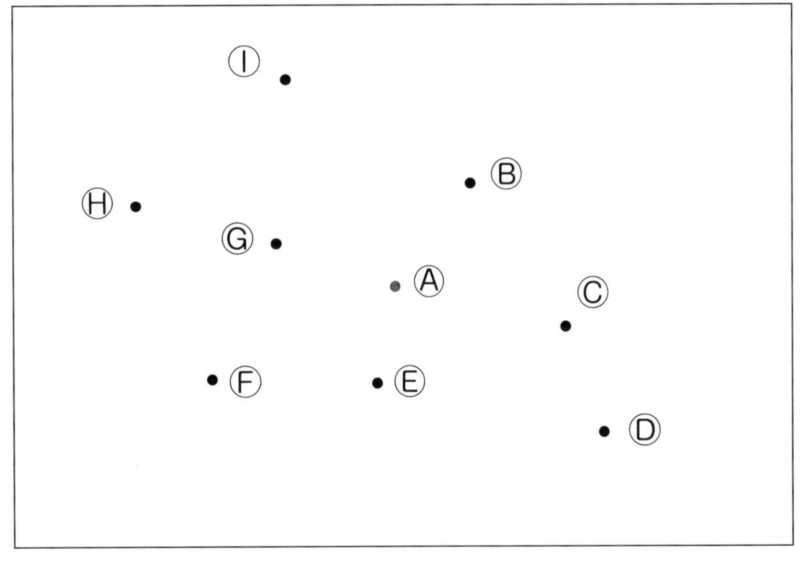

Ⓘ
Ⓑ
Ⓗ
Ⓖ
Ⓐ
Ⓒ
Ⓕ
Ⓔ
Ⓓ

① Which point from point Ⓐ has the same length from point Ⓐ to point Ⓑ?

② Point Ⓐ to some points has the longer length than the length from point Ⓐ to point Ⓘ. How many of them are there?

2 Spheres

1 Let's try to explore what kind of shapes can be seen when the following objects are viewed from the side or from the top.

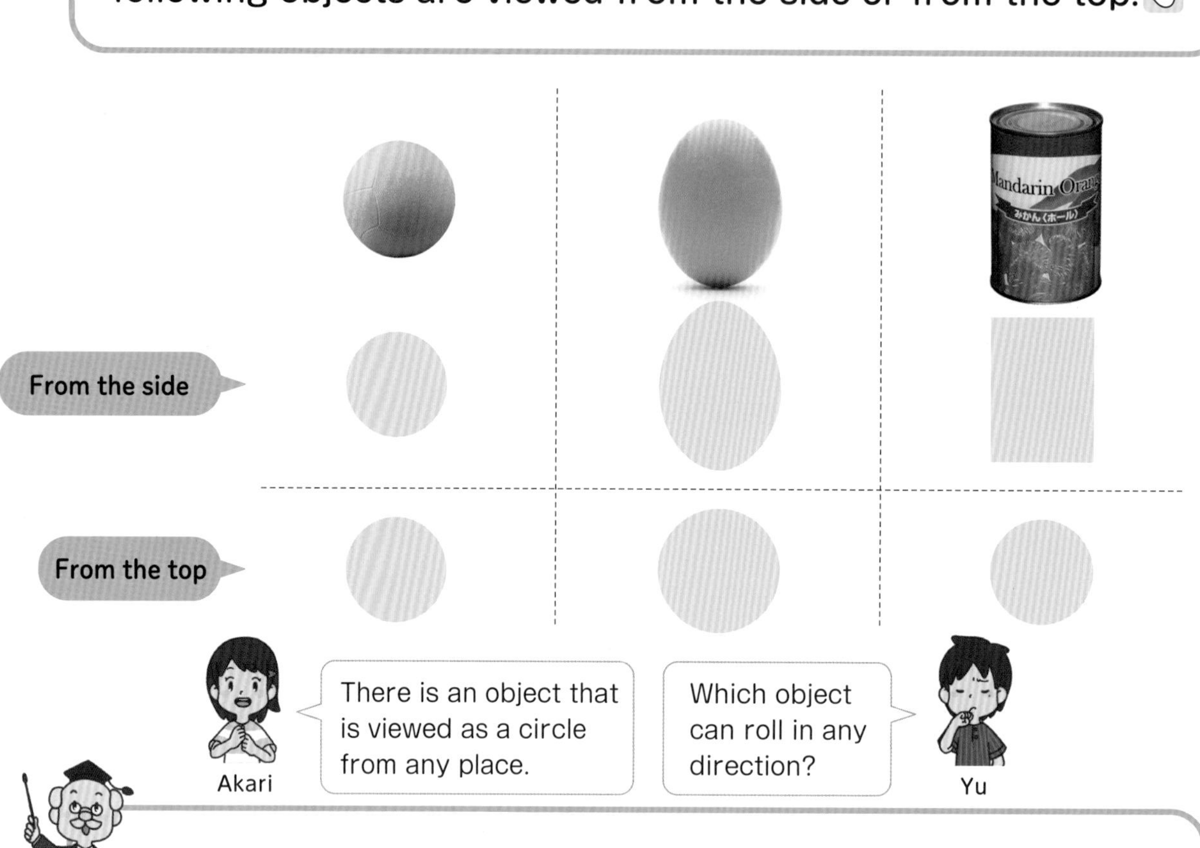

Mandarin Orange
From the side
From the top
There is an object that is viewed as a circle from any place.
Akari
Which object can roll in any direction?
Yu

A shape that looks like a circle from any direction is called a **sphere**.

\ Want to explore /

Purpose What properties does a sphere have?

1 What kind of shape is the cross-section of a sphere? Where should we cut to make the largest cross-section of a sphere? ▷

All cross-sections of a sphere are circles. The largest cross-section of a sphere is found when we cut it exactly in half.

When a sphere is cut in half, the center, the radius, and the diameter of the cross-section are called the **center**, **radius**, and **diameter** of the sphere respectively.

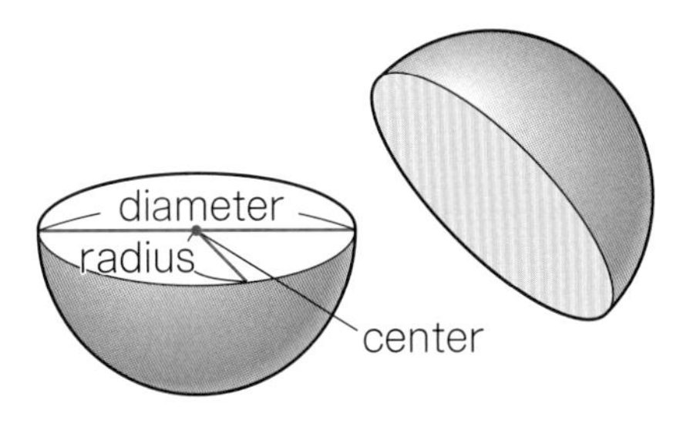

2 How can we find out the diameter of a sphere?

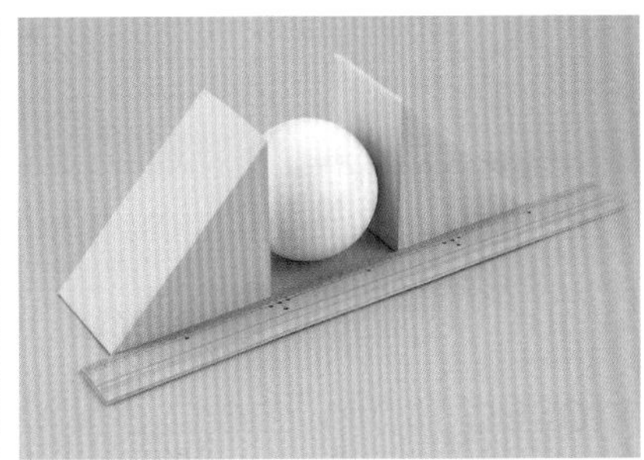

Way to see and think

We also explored the diameter of a circle in the same way.

3 Let's look for objects with the shape of a sphere in our surroundings.

(Higashikawa City, Kamikawa County, Hokkaido Prefecture)

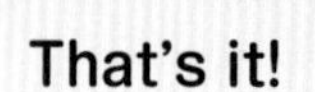

That's it! Why are manhole covers circular?

Haruto: I see manholes when I am walking on the streets.

Akari: Most of the manhole covers are circular. Is there any reason for this?

Let's think about the reasons why manhole covers are circular.

(Takatsuki City, Osaka Pref.)

(Sapporo City, Hokkaido Pref.)

① Let's discuss what happens when the covers are removed from the manholes.

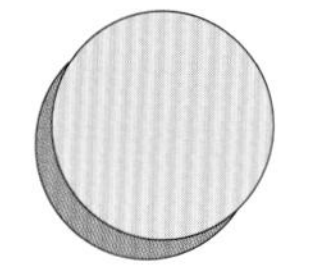

Yu: The longest line with a circle is the diameter, so...

Sara: The longest line within a quadrilateral is...

② Let's investigate about other shapes of manhole covers in our country.

Haruto: I have seen one in a quadrilateral shape.

(Kobe City, Hyogo Pref.)

The quadrilateral manholes are also called "handholes." People cannot enter most of them.

CAN What can you do?

☐ We understand about the diameter and radius of a circle. → p.104, p.109

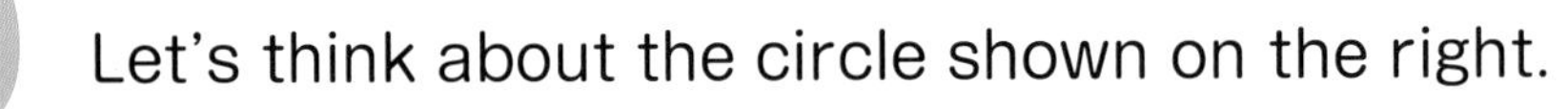

1 Let's think about the circle shown on the right.

① What is point Ⓐ called?

② What are the straight lines Ⓑ and Ⓒ called?

③ Write the word that apply to the following ☐, about the relationship between lines Ⓑ and Ⓒ.

The length of Ⓑ is ☐ the length of Ⓒ.

☐ We can draw circles. → pp.106 ～ 107

2 Let's draw the following circles.

① a circle with a diameter of 4cm

② a circle with a radius of 4cm

③ a circle with a radius of 6cm

☐ We can use the compass. → p.111

3 Let's use a compass to compare the length of the following straight lines, and arrange symbols Ⓐ, Ⓑ, and Ⓒ from the longest to the shortest.

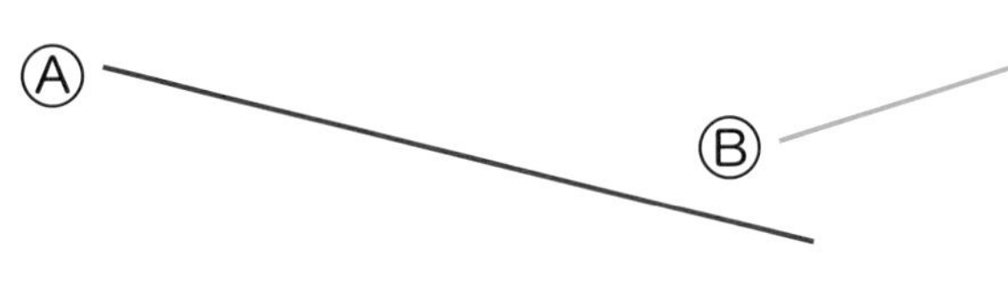

☐ We understand about spheres. → p.113

4 5 balls with a radius of 4cm are placed in the box as shown on the right. Let's find out the length, width, and height of the box.

height

width

length

Supplementary Problems → p.138

Which "Way to See and Think Monsters" did you find in "7 Circles and Spheres"?

I found "Why" when I was examining whether a certain shape is a circle or not.

Yu

I found other monsters too!

Sara

Usefulness and Efficiency of Learning

1 The figure on the right shows a circle that fits exactly in a square. Let's explore the radius of this circle and draw a circle of the same size.

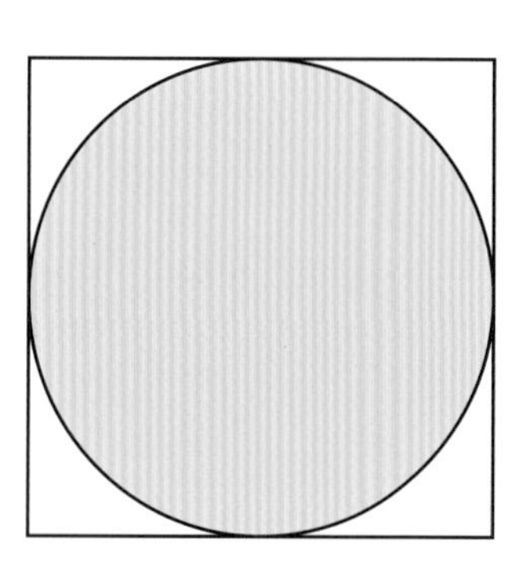

2 Which length is longer, the length surrounding the rectangle or that of the square? Let's use a compass and the straight lines drawn below.

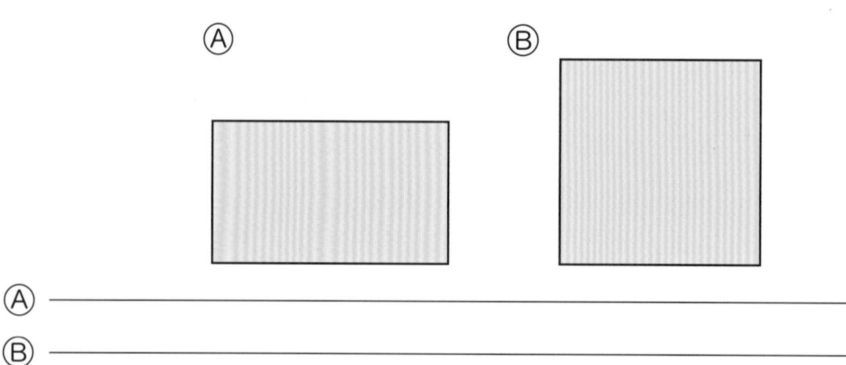

Ⓐ ______________________________

Ⓑ ______________________________

3 There was a broken plate in the treasure box. The original shape was a circle. Let's draw the original circle.

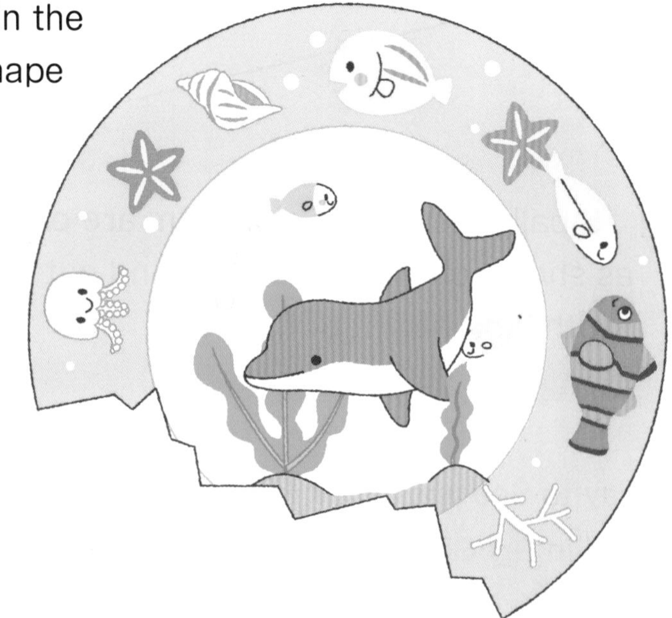

Let's find the center of the circle by tracing the plate on the right on a piece of tracing paper.

With the Way to See and Think Monsters...

Let's Reflect!

Let's reflect on which monster you used while learning "7 Circles and Spheres."

Why

By understanding the properties of a circle, we could explain why some round shapes cannot be called a circle.

① Let's explain why the following shape cannot be called a circle.

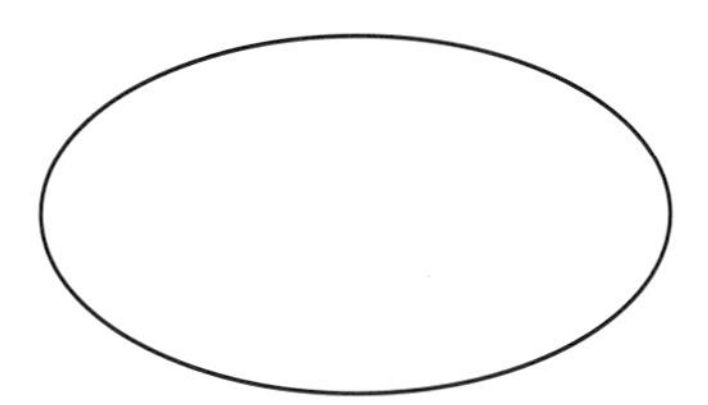

Circle is a round shape of which the lengths of all radius are the [ ].
The shape on the left is not so.

Same Way

Both a circle and a sphere has a center, a diameter, and a radius. The length of the diameter is always double the length of the radius.

② What is the similarity between a circle and a sphere?

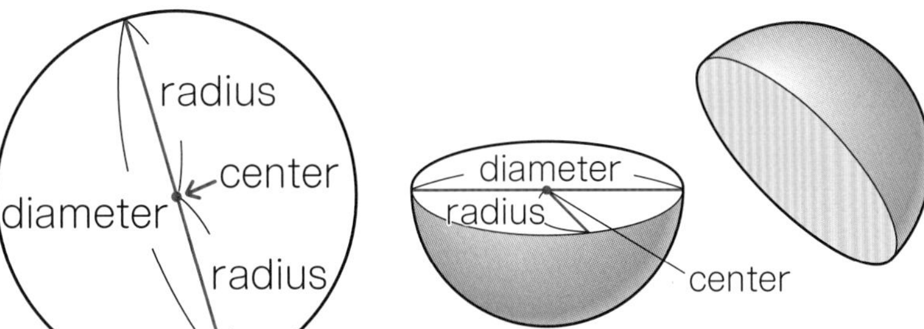

In both a circle and a sphere, the length of the radius is the [ ] length of the diameter.

Solve the ?

By drawing a shape with many points at the same distance from one point, it became a circle. We could draw a circle easily by using a compass.

Want to Connect

Are there other usages for a compass besides from drawing circles?

Find the ?

Is it divisible?

Let's think about how to calculate when the number cannot be divided exactly.

Division with Remainders

Let's think about the meaning of remainder in division.

Way to see and think

Let's arrange in sets of 6.

If I put 6 oranges in each bag...

How many bags can be filled?

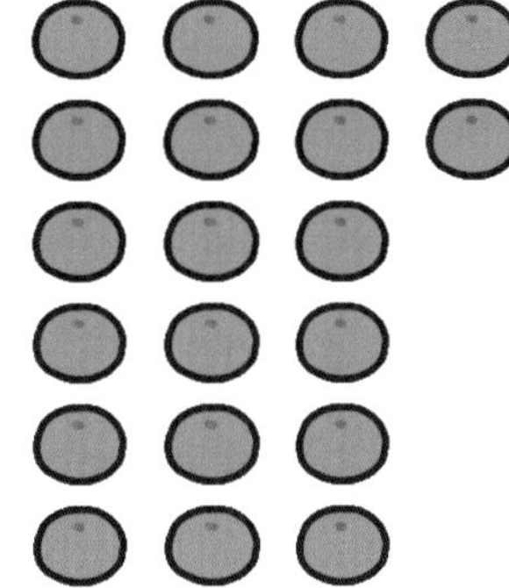

1 Division with remainders

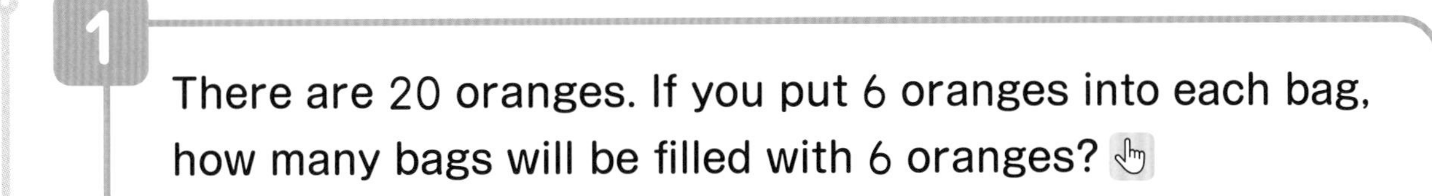

1 There are 20 oranges. If you put 6 oranges into each bag, how many bags will be filled with 6 oranges?

① Let's write a math expression.

② Look at the picture above. How many bags will be filled with oranges? How many oranges will remain?

③ Let's think about how to calculate.

Sara: Can I find the answer by using the multiplication table?

Haruto: Is there the multiplication table for "6 ×□= 20?"

Want to think

? Purpose Let's think about how to calculate 20 ÷ 6.

By using the multiplication table, it is shown as the following.

	Number for each		Number of units		Total number	
1 bag	6	×	1	=	6	14 oranges remain
2 bags	6	×	2	=	12	8 oranges remain
3 bags	6	×	**3**	=	**18**	**2** oranges remain
4 bags	6	×	4	=	24	4 oranges short

Way to see and think

Putting various numbers in the "number of units."

There are 20 oranges. If you put 6 oranges into each bag, there will be 3 bags and 2 remaining oranges. This can be written as follows.

$20 \div 6 = 3$ remainder 2

Answer: 3 bags and 2 oranges remain

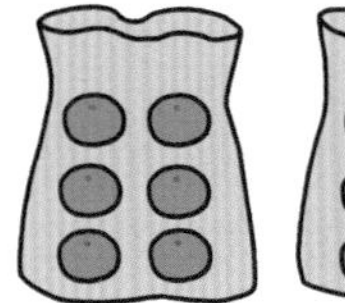
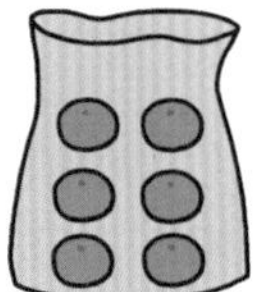

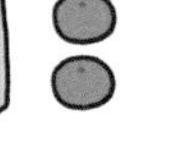

remainder

When there is a **remainder** after dividing, like $20 \div 6$, the division expression is called **not divisible**. When there is no remainder, like $18 \div 6$ and $24 \div 6$, the division expression is called **divisible**.

Summary

Even when the dividend is not divisible, like $20 \div 6$, the answer to a division problem can be found using the multiplication table.

1 Let's calculate the following.

① $11 \div 2$ ② $48 \div 7$ ③ $17 \div 3$ ④ $65 \div 8$

2 There are 34 cards. If 6 children get the same number of cards, how many cards will each child get? How many cards will remain?

I wonder if there are times when the division is not divisible in the case to find out "Number for each."
Akari

? Is there any rule about the number of the remainder?

2

There are 23 chestnuts. If you put 4 chestnuts into each bag, how many bags will be filled? How many chestnuts will remain?

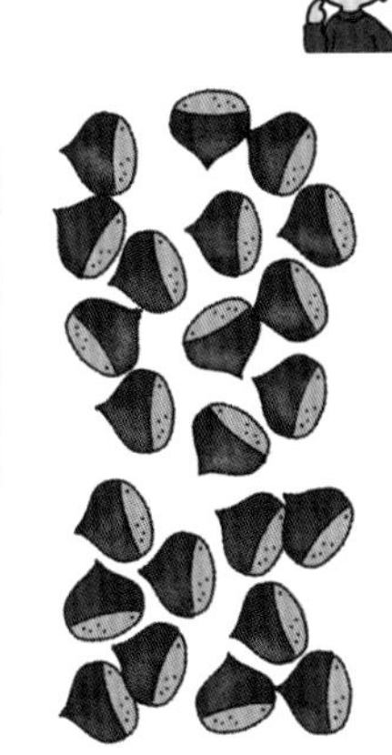

① Let's write a math expression.

□ ÷ □

② Let's discuss the ideas of Yu and Sara.

Yu's idea

Number for each		Number of units		Total number	
4	×	3	=	12	11 remainders
4	×	4	=	16	7 remainders

$23 \div 4 = 4$ remainder 7

Answer: 4 bags and 7 chestnuts remain

Sara's idea

Number for each		Number of units		Total number	
4	×	4	=	16	7 remainders
4	×	5	=	20	3 remainders
4	×	6	=	24	1 short

$23 \div 4 = 5$ remainder 3

Answer: 5 bags and 3 chestnuts remain

\ Want to explore /

In division, what relationship is there between the divisor and the size of the remainder?

1 Divisions, wherein the divisor is 4, are lined up on the right. Let's fill in the □ with numbers and examine the relationship between the divisor and the size of the remainder. Then, let's explain what you found out about the remainders.

Dividend		Divisor		Answer		Remainder
12	÷	4	=	3		
11	÷	4	=	2	remainder	3
10	÷	4	=	2	remainder	2
9	÷	4	=	2	remainder	1
8	÷	4	=	2		
7	÷	4	=	1	remainder	□
6	÷	4	=	1	remainder	□
5	÷	4	=	1	remainder	□
4	÷	4	=	1		
3	÷	4	=	□	remainder	□
2	÷	4	=	□	remainder	□
1	÷	4	=	□	remainder	□

What does "remainder 0" mean?

Haruto

Way to see and think

When the dividend is changed to consecutive numbers, the relationship between the answer and the remainder is made clear.

Summary

The remainder in division should always be smaller than the divisor.

2 Let's calculate the followings, paying attention to the size of the remainder.

① $7 \div 2$　② $10 \div 3$　③ $14 \div 4$

④ $38 \div 7$　⑤ $43 \div 5$　⑥ $58 \div 6$

3 There are 41 jellies. If 5 children get the same number of jellies, how many jellies will each child get? How many jellies will remain?

Way to see and think

Is the remainder smaller than the divisor ?

? Is there a way to confirm whether the answer or the remainder is correct?

3

There are 26 candies. If you put 8 candies into each bag, how many bags will be filled? How many candies will remain?

① Let's fill in the ☐ with numbers.

☐ ÷ ☐ = ☐ remainder ☐

Total number | Number for each | Number of units | Remainder

Answer: ☐ bags and ☐ candies remain.

＼ Want to know ／

Purpose How can we confirm the answer of division?

② The answer to the division problem in ① can be confirmed by the following operation. Let's think about the reason using the diagram on the right.

8 × 3 + 2 = ☐

Number for each | Number of units | Remainder | Total number

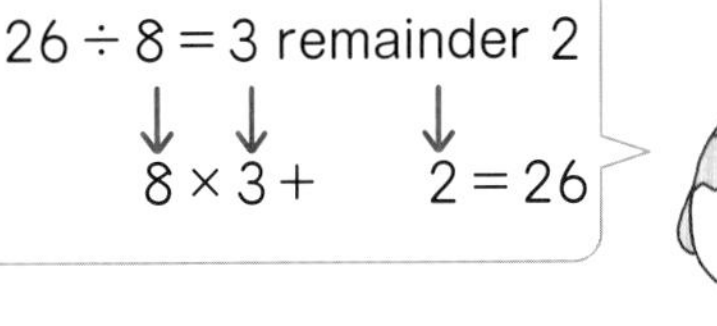

1 Let's confirm the answers of the following. If the answer is not correct, write the correct answer.

① 29 ÷ 4 = 7 remainder 1　② 19 ÷ 4 = 5 remainder 1

③ 34 ÷ 7 = 4 remainder 6　④ 50 ÷ 6 = 7 remainder 8

2 Let's calculate the following and confirm the answers.

① 7 ÷ 4　② 47 ÷ 9　③ 50 ÷ 7

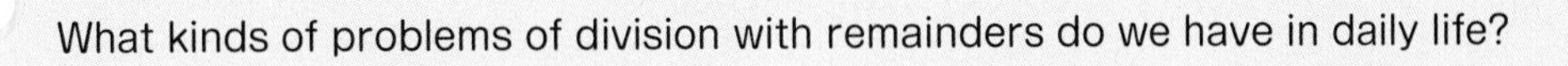

? What kinds of problems of division with remainders do we have in daily life?

2 Various Problems

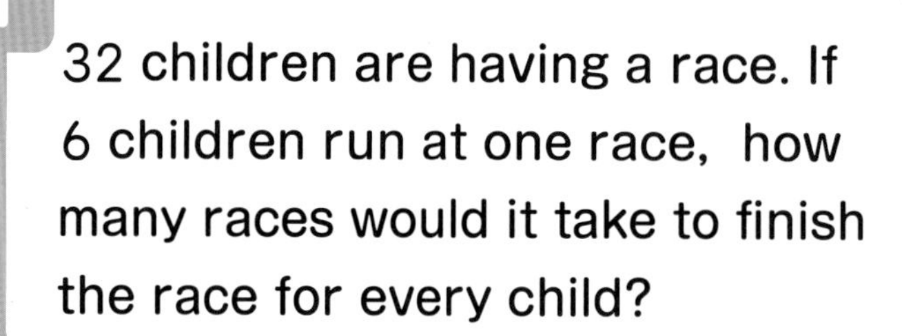

1 32 children are having a race. If 6 children run at one race, how many races would it take to finish the race for every child?

① Let's write a math expression. [ ]

② If 6 children run at one race, how many groups will there be and how many children will remain?

If it is 5 groups, $6 \times 5 = 30$, so 2 children will remain.

Then there would be a race with only 2 children...

How about making 6 groups?

How about putting 7 children for one race?

\ Want to think /

Purpose How can we deal with the childen remaining?
Yu

③ If we make a group of 6 children and a group of 7 children, how many groups will there be respectively? How many races do we need to have all groups run?

Way to see and think

Can we include the children remaining to other groups?

1 There are 29 children. If there is a bench that 9 children can sit at once, how many benches are needed for all children to be seated?

All children should be able to sit, so...

2 There are 62 oranges. If you put 7 into each box, how many boxes will there be?

We can't fill one box with the number of oranges left.

CAN What can you do?

☐ We can solve divisions with remainders. → pp.120 ~ 122

1 Let's calculate the following.

① 33 ÷ 8　② 48 ÷ 5　③ 17 ÷ 4　④ 26 ÷ 7
⑤ 56 ÷ 9　⑥ 43 ÷ 6　⑦ 13 ÷ 2　⑧ 39 ÷ 7
⑨ 74 ÷ 9　⑩ 70 ÷ 8　⑪ 7 ÷ 6　⑫ 4 ÷ 9

☐ We understand the relationship between the divisor and the size of the remainder. → p.122

2 Let's find the mistakes in the following divisions. Write the correct answers in the ☐.

① 37 ÷ 5 = 8 remainder 3　② 28 ÷ 3 = 8 remainder 4

☐ We can confirm the answer in a division. → p.123

3 Let's confirm the answers in the following divisions.

① 19 ÷ 3 = 6 remainder 1
3 × ☐ + ☐ = ☐

② 31 ÷ 4 = 7 remainder 3
☐ × ☐ + ☐ = ☐

☐ We can make a division expression and find out the answer. → p.124

4 There were 66 cards. let's answer the following.

① If 9 children get the same number of cards, how many cards will each child get? How many cards will remain?

② If each child gets 9 cards, how many children can share the cards? How many cards will remain?

Supplementary Problems → p.139

Which "Way to See and Think Monsters" did you find in "8 Division with Remainders"?

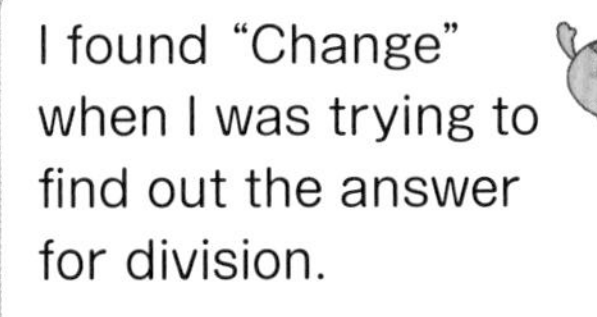

I found "Change" when I was trying to find out the answer for division.

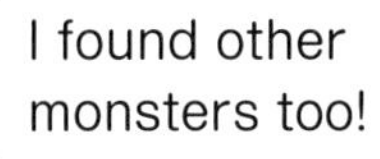

I found other monsters too!
Yu

Usefulness and Efficiency of Learning

1 You got 46 persimmons. You would like to divide them equally among 6 people. Let's answer the following.

① How many persimmons will each person get? How many persimmons will remain?

② How many more persimmons do you need if you want to give 8 persimmons to each person?

2 There are 29 balls to be carried. If you carry 3 balls at a time, how many times do you need to carry the balls?

3 There are 61 cookies. Let's answer the following.

① If you put 7 cookies to each bag, how many bags will be filled? How many cookies will remain?

② If you put 7 cookies or 8 cookies into each bag without a remainder, how many bags of 7 cookies and bags of 8 cookies will be filled?

That's it!

Division algorithm in vertical form

Advanced
4th grade

As well as in addition and subtraction, we can make division algorithm in vertical form as shown below.

6) 45 How to write:

45 ▶) 45 ▶ $\overline{)45}$ ▶ 6 $\overline{)45}$

Division algorithm for 45 ÷ 6 in vertical form

$6\overline{)4\ 5}$ (with □ above) →

divide

$$\begin{array}{r} 7 \\ 6\overline{)4\ 5} \end{array}$$

Write 7 above the digit in the ones place of 45.

→ **multiply**

$$\begin{array}{r} 7 \\ 6\overline{)4\ 5} \\ 4\ 2 \end{array}$$

Write 42, which is the answer to 6 × 7, below 45. Align the digits of the numbers according to their places.

→ **subtract**

$$\begin{array}{r} 7 \\ 6\overline{)4\ 5} \\ 4\ 2 \\ \hline 3 \end{array}$$

answer: 7
remainder: 3

The result when 42 is subtracted from 45 is 3. The remainder is 3. This is smaller than the divisor 6.

With the Way to See and Think Monsters...

Let's Reflect!

Let's reflect on which monster you used while learning "8 Division with Remainders."

Change

We changed the numbers for "Number of units" in various ways to find out which answer fits the problem.

① There are 20 oranges. If you put 6 oranges into each bag, how many bags will be filled?

	Number for each		Number of units		Total number	
1 bag	6	×	1	=	6	14 oranges remain
2 bags	6	×	2	=	12	8 oranges remain
3 bags	6	×	**3**	=	**18**	**2** oranges remain
4 bags	6	×	4	=	24	4 oranges short

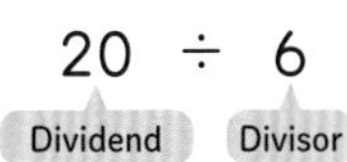

By applying various numbers in the "Number of unit", we tried to find the case where the remainder gets smaller than the divisor.

Yu

Why

We thought about the reason depending on the purpose of what we want to know.

② There are 28 children in Eita's class. If the class is divided into groups of 5 children, how many groups will be formed? How many children will remain? What can we do to have no children remaining?

If we make the groups of 5, ☐ children will remain. By putting one each to the 5 groups, we can make ☐ groups of six.

Akari

Solve the ?

In cases of division that is not divisible, by thinking about the remainder and using the multiplication table, we could find out the answer.

Sara

Want to Connect

Can we do division of larger numbers?

Haruto

More Math!

[Supplementary Problems]

[Let's deepen.]

[Answers]

Multiplication

→ pp.12 ～ 25

1 Let's fill in each □ with a number.

① $7 \times 3 = 3 \times \square$　　② $8 \times 6 = \square \times 8$

2 Let's fill in each □ with a number.

① 6×5 is larger than 6×4 by □ .

② 9×4 is larger than $9 \times \square$ by 9.

③ 8×7 is smaller than 8×8 by □ .

④ 3×5 is smaller than $3 \times \square$ by 3.

⑤ $4 \times 9 = 4 \times \square + 4$

⑥ $7 \times 6 = 7 \times 5 + \square$

⑦ $5 \times 7 = 5 \times 8 - \square$

⑧ $9 \times 8 = 9 \times \square - 9$

3 Let's fill in each □ with a number.

① 8×5 <
$3 \times 5 = \square$
$\square \times 5 = \square$
Total □

② 9×7 <
$9 \times 2 = \square$
$9 \times \square = \square$
Total □

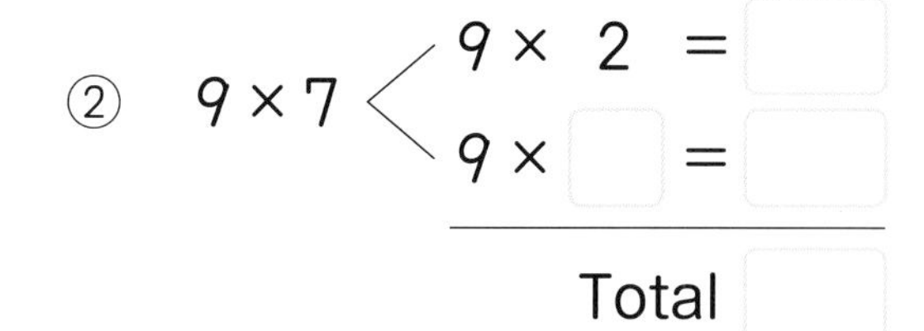

4 Let's fill in each □ with a number.

① $(3 \times 2) \times 2 = \square \times 2$
$= \square$

② $3 \times (2 \times 2) = 3 \times \square$
$= \square$

③ $(3 \times 2) \times 2 = \square \times (2 \times 2)$

5 Let's calculate the following.

① 2×0　② 7×0　③ 0×5　④ 9×0

⑤ 0×3　⑥ 0×8　⑦ 3×10　⑧ 7×10

⑨ 9×10　⑩ 10×6　⑪ 10×8　⑫ 10×5

6 There are 4 boxes of cookies that contain 10 cookies each. How many cookies are there altogether?

2 Time and Duration (1)

→ pp. 26 ~ 35

1 Let's find out the following time.

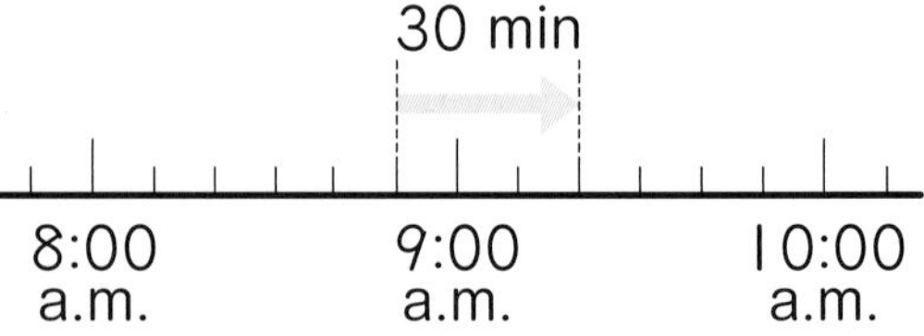

① The time 30 minutes past 8:50 a.m.

② The time 1 hour 20 minutes past 10:30 a.m,

③ The time 2 hours 20 minutes past 6:45 p.m.

2 Let's find out the following duration.

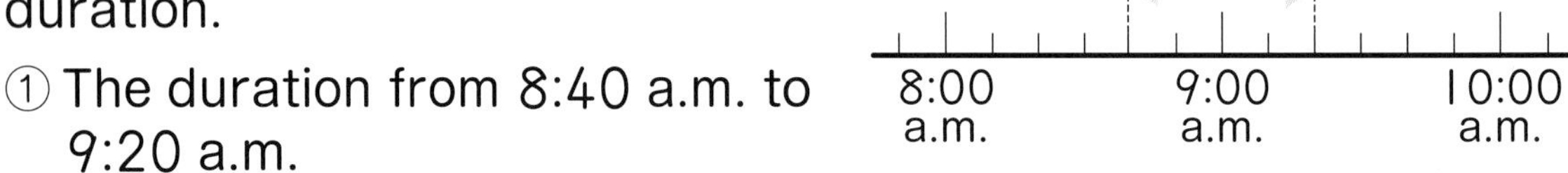

① The duration from 8:40 a.m. to 9:20 a.m.

② The duration from 1:50 p.m. to 2:40 p.m.

③ The duration from 4:45 p.m. to 7:15 p.m.

3 Let's find out the following time.

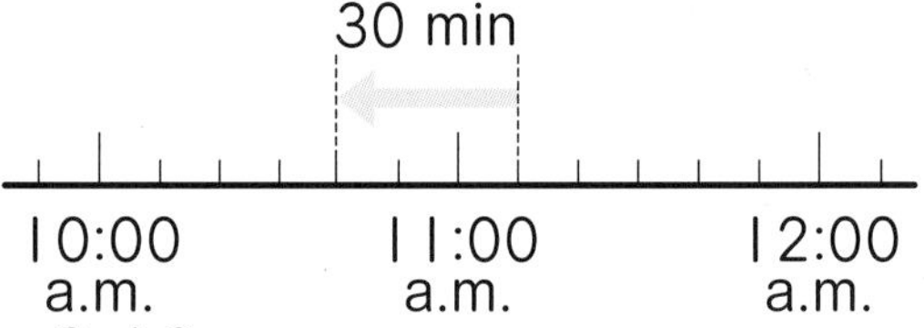

① The time 30 minutes before 11:10 a.m.

② The time 1 hour 40 minutes before 9:40 a.m.

③ The time 2 hours 30 minutes before 5:15 p.m.

4 Let's find out the following time and duration.

① The time 30 minutes before 8:50.

② The duration from 1:20 to 6:50.

③ The time 40 minutes past 1:40.

④ The time 1 hour 40 minutes past 4:10.

	8	50
−		30

	6	50
−	1	20

	1	40
+		40

	4	10
+	1	40

5 Let's fill in the ☐ with numbers.

① 1 minute 25 seconds = ☐ seconds

② 1 minute 52 seconds = ☐ seconds

③ 97 seconds = ☐ minutes ☐ seconds

④ 108 seconds = ☐ minutes ☐ seconds

3 Division

→ pp.36 ～ 53

1 If you divide 18 chocolates equally among 3 children, how many chocolates would each child get? Let's write a math expression. ☐

2 If you divide 15 dL of milk equally into 5 cups, how many dL of milk would each cup have? Let's write a math expression.

☐

3 Which row of the multiplication table can be used to find out the answer to the following divisions? Let's find out the answer.

① $4 \div 2$ ② $28 \div 7$ ③ $56 \div 8$

④ $24 \div 4$ ⑤ $48 \div 6$ ⑥ $63 \div 9$

4 You are going to make a math problem for each expression. Let's fill in the ☐ with numbers.

① $24 \div 3$

If you divide ☐ sheets of colored papers equally among ☐ children, how many sheets would each child get?

② $30 \div 5$

If you cut a ☐ cm tape equally into ☐ , how long would each tape be?

5 Let's calculate the following.

① $16 \div 8$ ② $14 \div 2$ ③ $27 \div 3$

④ $35 \div 5$ ⑤ $24 \div 6$ ⑥ $28 \div 4$

6 There are 18 candies and each child gets 3 candies. Let's think about how many children can share the candies.

① Let's write a math expression.

② Which row in the multiplication table can be used to find out the answer to the division you made in ①?

③ How many children can share the candies?

7 There are 20 dL of juice. If you pour 4 dL to each bottle, how many bottles will be filled?

8 Let's calculate the following.

① $14 \div 7$ ② $18 \div 9$ ③ $25 \div 5$ ④ $36 \div 4$ ⑤ $21 \div 3$

⑥ $16 \div 2$ ⑦ $56 \div 7$ ⑧ $42 \div 6$ ⑨ $54 \div 9$ ⑩ $72 \div 8$

9 You are going to make a math problem for $27 \div 9$. Let's fill in the □ with numbers.

> There are □ cookies. If each child gets □ cookies, how many children can share the cookies?

10 There are 30 cards. Let's think about how to share them.

① If 5 children get the same number of cards, how many cards does each child get?

② If each child gets 6 cards, how many children can share the cards?

11 Let's think about how to cut a 24 cm ribbon.

① If you cut it every 6cm, how many ribbons can you get?

② If you cut it equally into 3, how long would each ribbon be?

12 Let's calculate the following.

① $8 \div 8$ ② $3 \div 3$ ③ $0 \div 7$ ④ $4 \div 1$

⑤ $20 \div 2$ ⑥ $60 \div 3$ ⑦ $44 \div 4$ ⑧ $22 \div 2$

Addition and Subtraction

→ pp.56 ~ 73

1 Let's calculate the following in vertical form.

① 352 + 416	② 316 + 253	③ 648 + 151
④ 652 + 107	⑤ 108 + 471	⑥ 306 + 401
⑦ 128 + 433	⑧ 516 + 248	⑨ 367 + 527
⑩ 247 + 236	⑪ 678 + 119	⑫ 365 + 308
⑬ 362 + 451	⑭ 671 + 275	⑮ 240 + 380
⑯ 693 + 237	⑰ 189 + 442	⑱ 736 + 89
⑲ 273 + 229	⑳ 415 + 387	㉑ 532 + 369
㉒ 656 + 144	㉓ 488 + 312	㉔ 334 + 68

2 There are 257 sheets of red paper and 163 sheets of blue paper. How many sheets of paper are there in total?

3 Let's calculate the following in vertical form.

① 873 − 241	② 659 − 448	③ 679 − 565
④ 689 − 525	⑤ 576 − 256	⑥ 816 − 511
⑦ 256 − 138	⑧ 674 − 405	⑨ 630 − 215
⑩ 816 − 332	⑪ 345 − 263	⑫ 718 − 631
⑬ 556 − 278	⑭ 322 − 199	⑮ 934 − 289
⑯ 311 − 163	⑰ 340 − 165	⑱ 614 − 58
⑲ 504 − 346	⑳ 906 − 438	㉑ 804 − 459
㉒ 604 − 206	㉓ 200 − 153	㉔ 500 − 45

4 Yumi has read 145 pages of a book with 240 pages. How many pages does she still need to read?

5 Let's calculate the following in vertical form.

① $685+536$ ② $483+517$ ③ $1327-943$

④ $1007-979$ ⑤ $3496+4207$ ⑥ $7382+2618$

⑦ $7134-3146$ ⑧ $7225-4627$ ⑨ $10000-3027$

6 Let's fill in each □ with a number.

① $258+98=258+\square-2$
$=\square-2$
$=\square$

② $406-197=406-\square+3$
$=\square+3$
$=\square$

③ $157+76+24$
$=157+(\square+24)$
$=157+\square$
$=\square$

7 Let's calculate the following in easier ways.

① $199+165$ ② $302-98$

③ $277+68+32$ ④ $43+188+57$

8 Let's calculate the following mentally.

① $17+8$ ② $56+16$ ③ $26-9$ ④ $82-24$

9 There are 416 children in Hiroka's school. 198 of them are boys. How many girls are there in her school?

10 165 red cosmoses bloomed. The number of white cosmoses bloomed is 15 more than that of red cosmoses. How many white cosmoses bloomed?

5 Tables and Graphs

→ pp.76 ～ 87

1 The table on the right shows the data about the vehicles that passed in front of the school. Let's change the characters "正" to numbers. Then, let's find the total number.

Vehicles that passed in front of the school

Kind	Number of vehicles	
Truck	正	
Car	正 下	
Bus	T	
Others	下	
Total		

2 The graph on the right shows the data about vehicles that passed in front of the station.

① How many vehicles does 1 cell represent?

② What kind of vehicle passed most frequently? How many were there?

③ How many trucks and buses were there respectively.

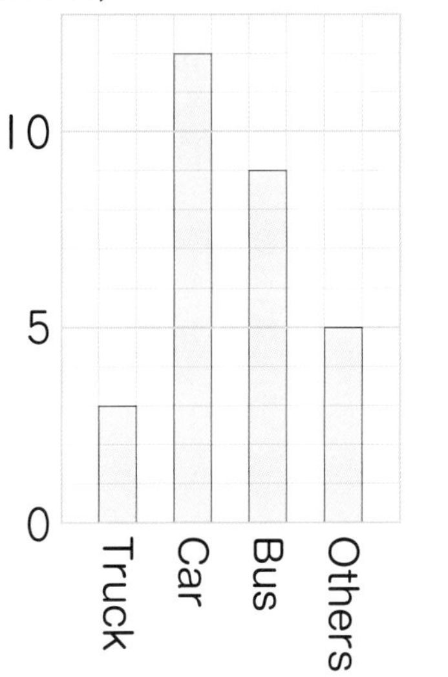

3 The bar graph on the right shows the number of empty cans collected by the children in each grade.

① How many cans does 1 cell represent?

② Which grade collected the most cans? How many did they collect?

③ Which grade collected more cans, 3rd or 6th, and by how many?

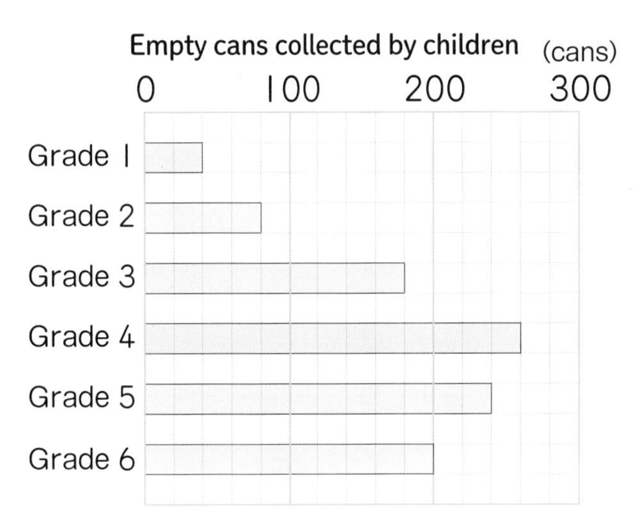

4 In the graphs below, let's tell the number that each bar represents.

①

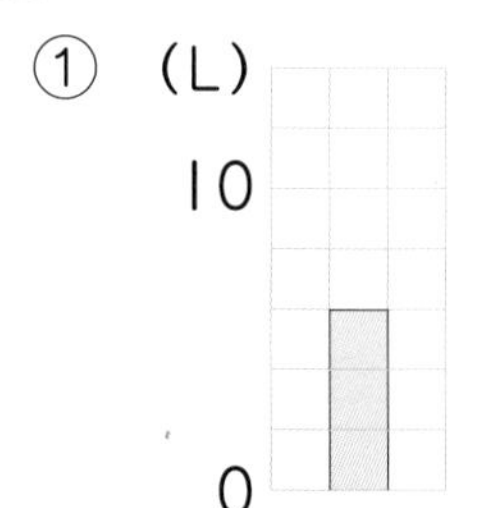

②

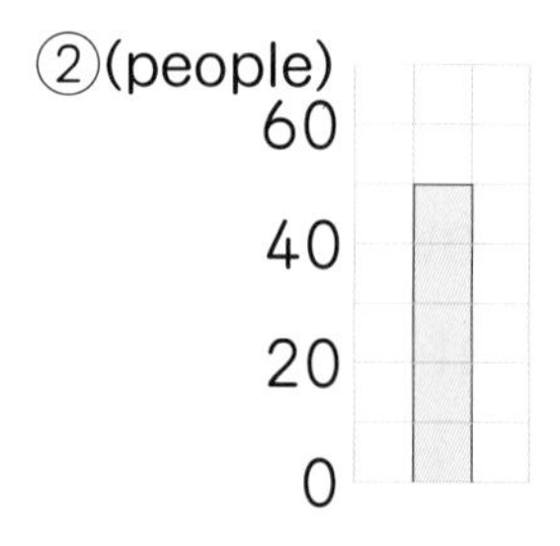

③ (m)

30

20

10

0

5 The table below shows the favorite fruits of children in Daichi's class. Let's draw a bar graph.

Favorite fruits

Fruit	Number of children
Apple	12
Grape	9
Strawberry	6
Orange	5
Others	4

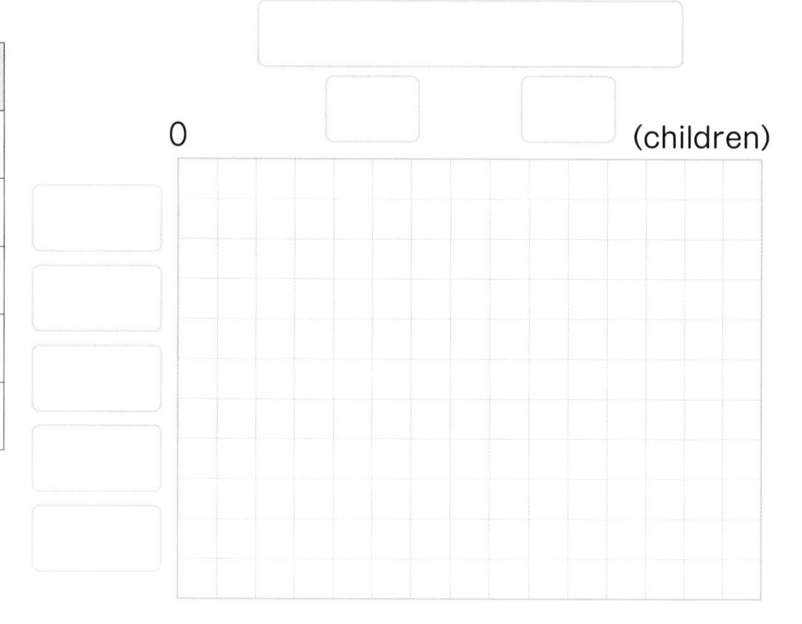

6 The following tables show the number of the 3rd grade children who borrowed each kind of books in April, May, and June. Let's organize the tables for each month into one table.

April

Kind	Number of books
Fiction	13
Biography	5
Picture	7
Others	4

May

Kind	Number of books
Fiction	18
Biography	12
Picture	9
Others	6

June

Kind	Number of books
Fiction	17
Biography	11
Picture	13
Others	8

Books borrowed by the 3rd grade children

(books)

Kind \ Month	April	May	June	Total
Fiction				
Biography				
Picture				
Others				
Total				

6 Length

→ pp.88 ～ 99

1 What is the length in m and cm of points ①, ②, ③, and ④ located on the tape measure below?

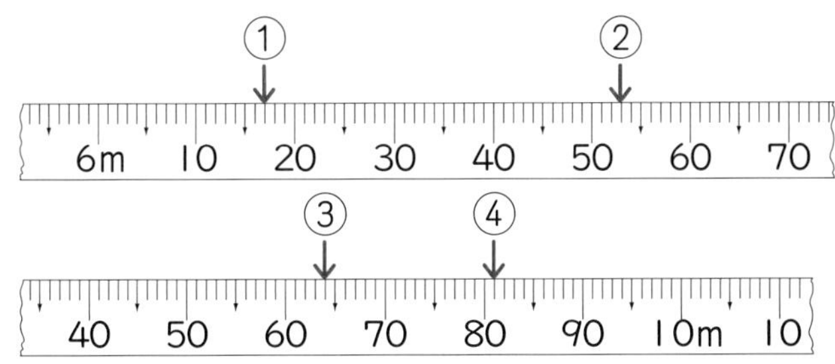

2 Let's fill in each ☐ with a number.

① 4000m = ☐ km ② 2850m = ☐ km ☐ m

③ 7km = ☐ m ④ 3km60m = ☐ m

3 Which is longer?

① 1km690m, 1800m ② 3km75m, 3065m

4 Look at the map on the right and answer the following.

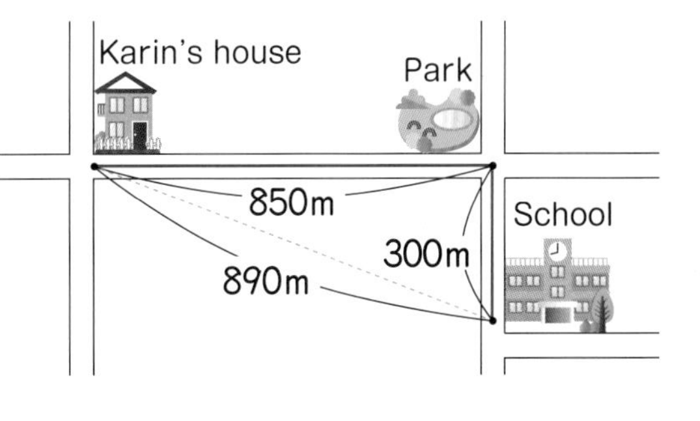

① What is the road distance in km and m from Karin's house to the school through the front of the park?

② What is the difference between the road distance and the distance from Karin's house to the school in m?

5 Let's calculate the following.

① 2km750m + 1km240m ② 2km380m + 1km640m

③ 2km480m − 1km260m ④ 3km80m − 1km630m

7 Circles and Spheres

→ pp.102 ~ 117

1 Let's write the appropriate words for Ⓐ, Ⓑ, and Ⓒ in the circle shown on the right.

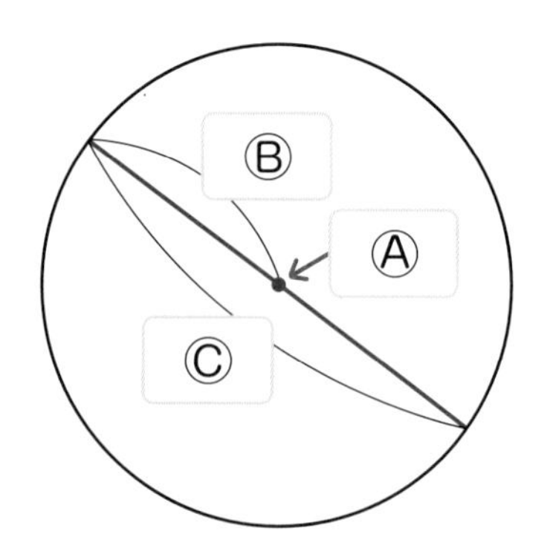

2 A circle is drawn to fit exactly inside a square with a side of 10 cm. How many cm is the radius?

3 Let's draw a circle with a radius of 4 cm.

4 Let's use a compass to compare the length of the following straight lines and arrange them from the longest to the shortest.

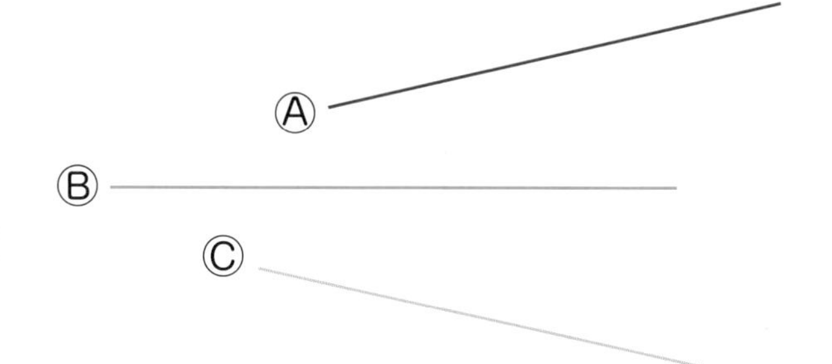

5 Each of the two circles shown on the right has a diameter of 10 cm. How many cm is the length of straight line AB?

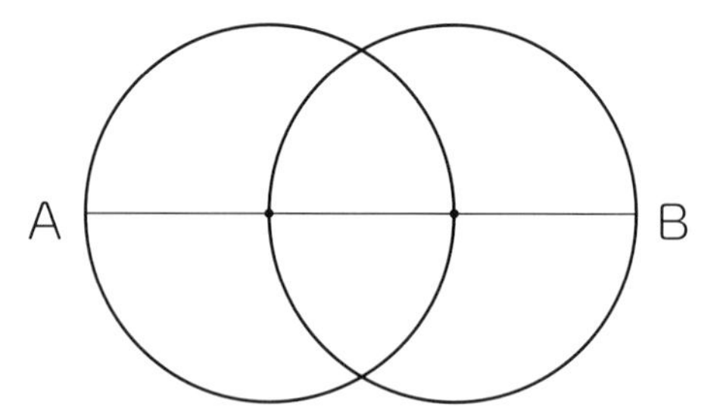

6 The three circles shown on the right have the same size. How many cm is the diameter of each circle when the length of the straight line AB is 28 cm?

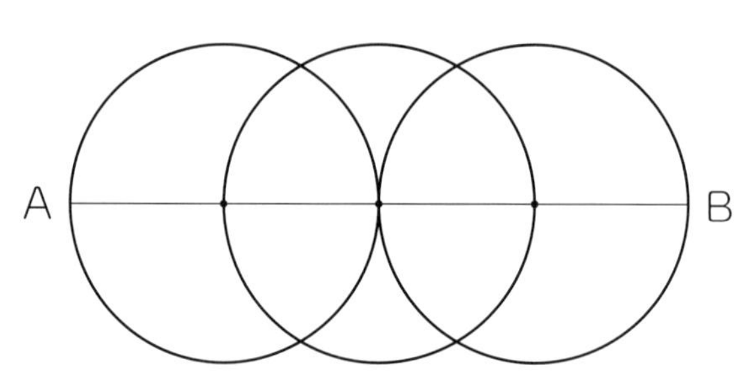

Division with Remainders

→ pp.118 ~ 127

1 Let's confirm the answers of the following. If the answer is not correct, write the correct answer.

① $30 \div 4 = 8$ remainder 2 ② $48 \div 7 = 7$ remainder 1

③ $29 \div 5 = 5$ remainder 4

2 Let's calculate the following and confirm the answers.

① $9 \div 4$ ② $8 \div 3$ ③ $46 \div 5$

④ $29 \div 6$ ⑤ $40 \div 9$ ⑥ $59 \div 7$

3 Let's think about how to distribute 46 sheets of drawing paper.

① If each child gets 6 sheets, how many children can share the drawing paper? How many sheets will remain?

② If 8 children get the same number of sheets, how many sheets does each one get? How many sheets will remain?

4 Let's think about how to cut a 50 cm ribbon.

① If you cut every 7 cm, how many ribbons can you get? How many cm will remain?

② If you want to get 8 ribbons of 7 cm, how many more cm of ribbon would you need?

5 There are 60 cakes that are to be placed on the plates. If 8 cakes can be placed on each plate, how many plates would be needed?

6 You are going to make packs of 6 eggs. There are 40 eggs. How many packs will be filled?

Let's play at the amusement park!

Taiga and his friends are going to the amusement park. Their plan is to will leave home at 9:40 a.m. and come back home at 3:00 p.m. They found out about the events and attractions as shown on page 141. Let's answer the following.

Can we make a plan at the amusement park?

Akari

① It takes 25 minutes from Taiga's house to the entrance of the amusement park. What time will they arrive at the entrance?

② Taiga and his friends are at the entrance. If they take a ride in the Ferris Wheel once and then play on the Go-cart once, and then ride the Teacups once, what would the total duration be?

You don' t need to take into account the duration to wait.

Sara

It takes 10 minutes from the entrance to the Ferris Wheel, and takes 10 minutes to ride the Ferris Wheel once. So, we can move to the next ride at the time 20 minutes past the time we arrive at the entrance.

③ Taiga wants to go to the Haunted House, and take lunch after that. To arrive at the restaurant at 11:50 a.m., by what time does he have to enter the Haunted House?

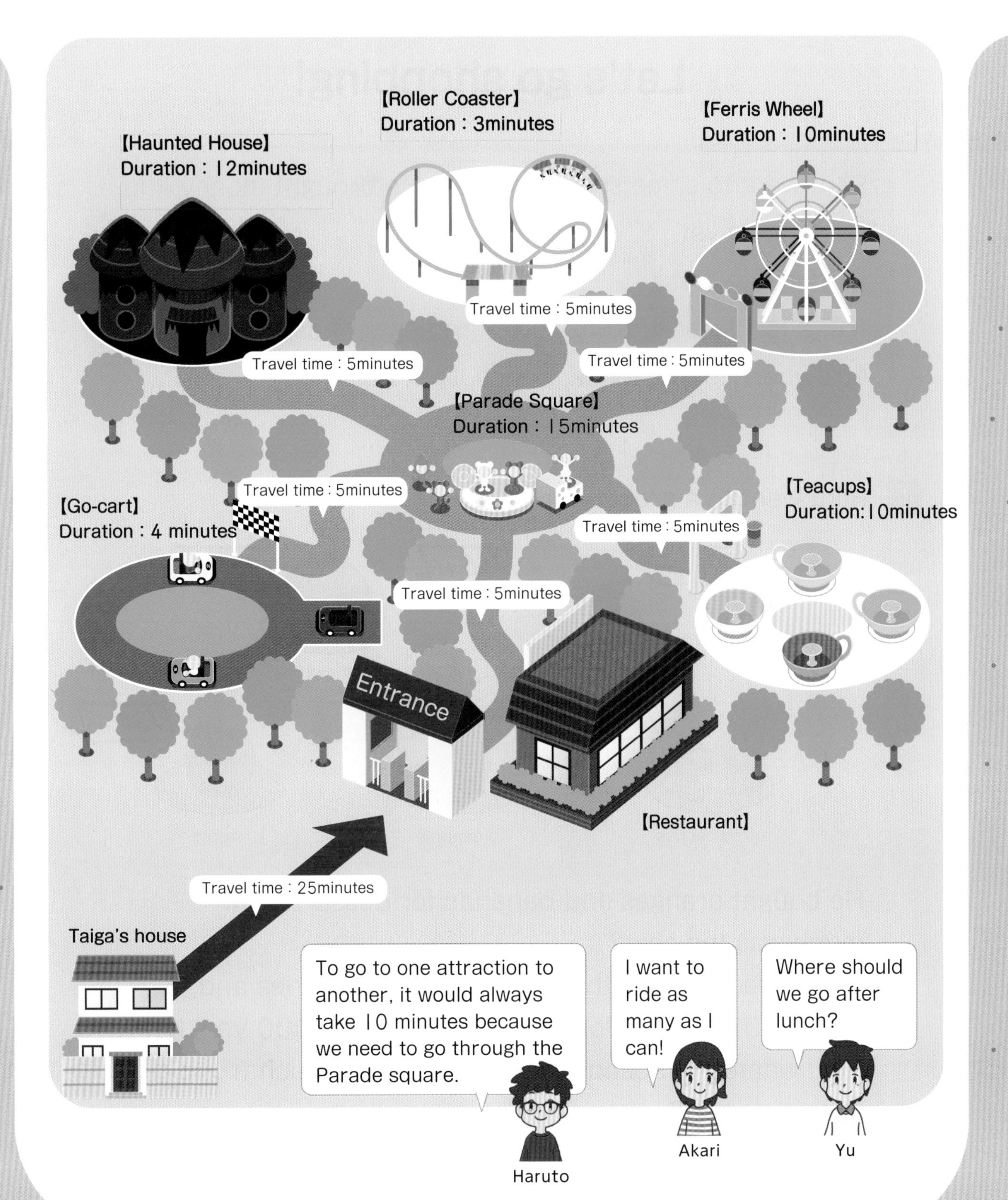
【Roller Coaster】
Duration : 3minutes
【Ferris Wheel】
Duration : 10minutes
【Haunted House】
Duration : 12minutes
Travel time : 5minutes
Travel time : 5minutes
Travel time : 5minutes
【Parade Square】
Duration : 15minutes
【Teacups】
Duration: 10minutes
Travel time : 5minutes
【Go-cart】
Duration : 4 minutes
Travel time : 5minutes
Travel time : 5minutes
Entrance
【Restaurant】
Travel time : 25minutes
Taiga's house
To go to one attraction to another, it would always take 10 minutes because we need to go through the Parade square.
I want to ride as many as I can!
Where should we go after lunch?
Haruto
Akari
Yu

Let's go shopping!

Riku went to some stores to shop. He brought money and a shopping list.

【Shopping List】

fruits for dessert, fruits for ingredients of cake, strawberries, notebook, pencil, light bulb

1 Riku went to a grocery store.

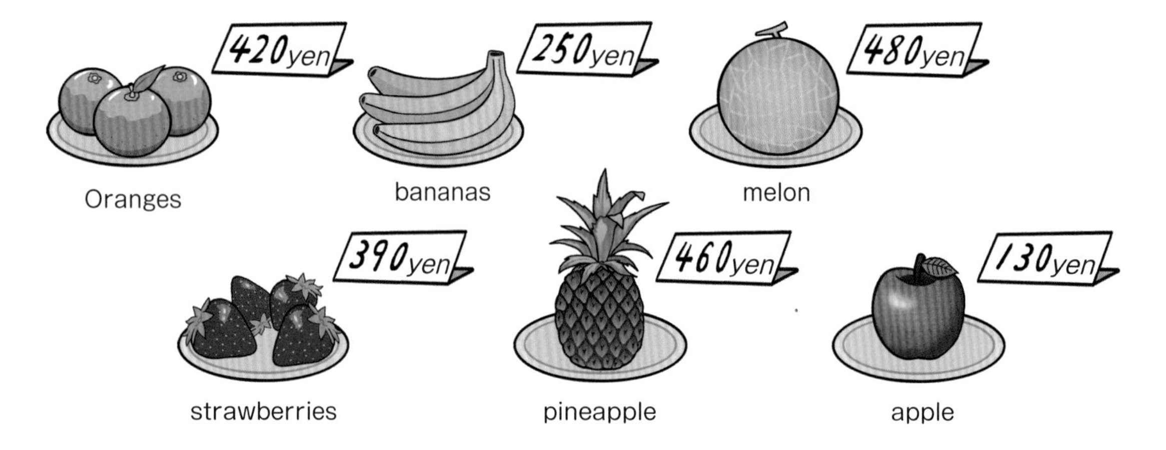

① He bought oranges and bananas for dessert. How much was the cost?

② The ingredients for the cake were strawberries and others. The budget for them was at most 1000 yen. If Riku wanted to spend exactly 1000 yen, which fruits should he buy?

2 Next, Riku went to a stationary shop. The amount of money left with him is shown on the right.
Riku bought two 60 yen pencils and a 200 yen notebook. Let's answer the following.

1000-yen bill 500-yen coin

100-yen coin 50-yen coin

10-yen coin

① How much was the total cost?

② If he had paid with a 500-yen coin, how much was the change?

③ With the amount of money left in his wallet, he paid to get the least total number of coins for change. Which coins did he use to pay?

3 After paying in the way in **2** ③, Riku finally went to the electrical shop. He bought a 630 yen light bulb.

① If he paid with 1000 yen, how much was the change?

② Riku paid to get the least total number of coins for change. How did he pay? How many coins are left in his wallet?

 (   ) coins    (   ) coins

 (   ) coins    (   ) coins

Is it a proper graph?

Yu investigated the number of typhoons which formed from 2016 to 2021. Let's answer the following.

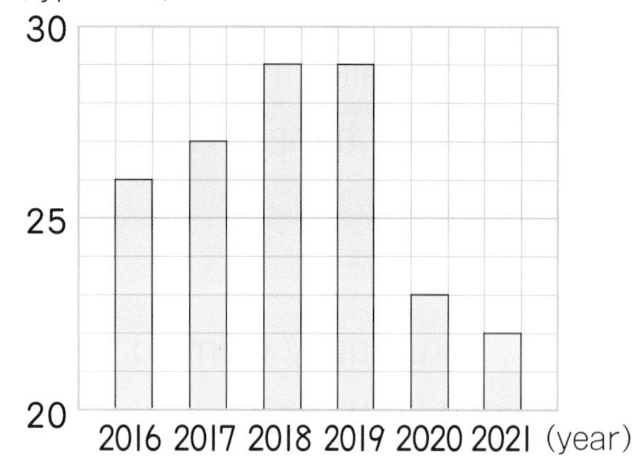

Akari: This graph shows that there were so many typhoons in 2018 and 2019.

Sara: The number of typhoons in 2016 is twice as many as in 2020, isn't it?

Haruto: In this graph, the units on the vertical axis don't start from 0 but from 20. Is this right?

① Three children are talking about the graph. Let's discuss how to think of their opinions.

② Let's draw the proper graph in the next page based on the table shown on the right.

Number of typhoons

Year	Number of typhoons
2016	26
2017	27
2018	29
2019	29
2020	23
2021	22

③ Let's compare Yu's graph and the graph you made in, and discuss what you found out.

Answers

[Supplementary Problems]

1 Multiplication

→ p.129

1 ① 7 ② 6

2 ① 6 ② 3 ③ 8 ④ 6
⑤ 8 ⑥ 7 ⑦ 5 ⑧ 9

3 ① 8×5 <
3 × 5 = 15
5 × 5 = 25
Total 40

② 9×7 <
9 × 2 = 18
9 × 5 = 45
Total 63

4 ① 6, 12 ② 4, 12 ③ 3

5 ① 0 ② 0 ③ 0 ④ 0
⑤ 0 ⑥ 0 ⑦ 30 ⑧ 70
⑨ 90 ⑩ 60 ⑪ 80 ⑫ 50

6 40cookies

2 Time and Duration (1)

→ p.130

1 ① 9:20a.m. ② 11:50a.m.
③ 9:05p.m.

2 ① 40 minutes ② 50 minutes
③ 2 hours 30 minutes

3 ① 10:40 a.m. ② 8:00 a.m.
③ 2:45 p.m.

4 ① 8:20 ② 5 hours 30 minutes
③ 2:20 ④ 5:50

5 ① 85 ② 112 ③ 1, 37 ④ 1, 48

3 Division

→ p.131

1 18 ÷ 3

2 15 ÷ 5

3 ① row of 2, 2 ② row of 7, 4
③ row of 8, 7 ④ row of 4, 6
⑤ row of 6, 8 ⑥ row of 9, 7

4 ① 24, 3 ② 30, 5

5 ① 2 ② 7 ③ 9 ④ 7 ⑤ 4 ⑥ 7

6 ① 18 ÷ 3 ② row of 3 ③ 6 children

7 5 bottles

8 ① 2 ② 2 ③ 5 ④ 9 ⑤ 7
⑥ 8 ⑦ 8 ⑧ 7 ⑨ 6 ⑩ 9

9 27, 9

10 ① 6 cards ② 5 children

11 ① 4 ribbons ② 8cm

12 ① 1 ② 1 ③ 0 ④ 4
⑤ 10 ⑥ 20 ⑦ 11 ⑧ 11

4 Addition and Subtraction

→ p.133

1 ① 768 ② 569 ③ 799 ④ 759
⑤ 579 ⑥ 707 ⑦ 561 ⑧ 764
⑨ 894 ⑩ 483 ⑪ 797 ⑫ 673
⑬ 813 ⑭ 946 ⑮ 620 ⑯ 930
⑰ 631 ⑱ 825 ⑲ 502 ⑳ 802
㉑ 901 ㉒ 800 ㉓ 800 ㉔ 402

2 420 sheets

3 ① 632 ② 211 ③ 114 ④ 164
⑤ 320 ⑥ 305 ⑦ 118 ⑧ 269
⑨ 415 ⑩ 484 ⑪ 82 ⑫ 87
⑬ 278 ⑭ 123 ⑮ 645 ⑯ 148
⑰ 175 ⑱ 556 ⑲ 158 ⑳ 468
㉑ 345 ㉒ 398 ㉓ 47 ㉔ 455

4 95 pages

5 ① 1221 ② 1000 ③ 384 ④ 28
⑤ 7703 ⑥ 10000 ⑦ 3988 ⑧ 2598
⑨ 6973

6 ① 100, 358, 356 ② 200, 206, 209
③ 76, 100, 257

7 ① 364 ② 204 ③ 377 ④ 288

8 ① 25 ② 72 ③ 17 ④ 58

9 218girls

10 180 white cosmoses

5 Tables and Graphs

→ p.135

1 Vehicles that passed in front of the school

Kind	Number of vehicles	
Truck	正	5
Car	正 下	8
Bus	丅	2
Others	下	3
Total	18	

2 ① 1 vehicle ② 12 cars
③ 3 trucks 9 buses

3 ① 20 cans ② 4th grade, 260 cans
③ 6th grade, 20 cans

4 ① 6L ② 50 children ③ 5m

5

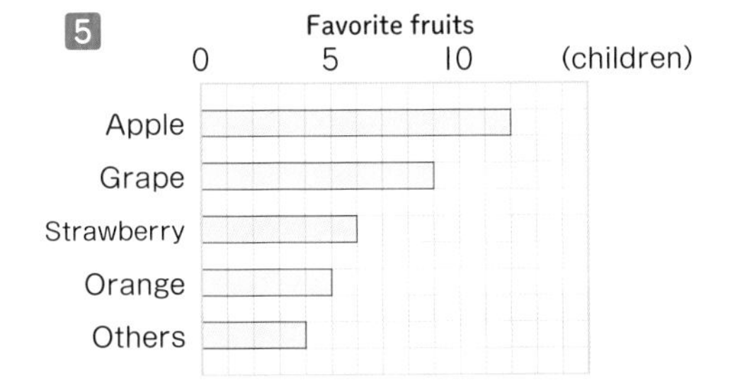

6 Books borrowed by the 3rd grade children (books)

Kind \ Month	April	May	June	Total
Fiction	13	18	17	48
Biography	5	12	11	28
Picture	7	9	13	29
Others	4	6	8	18
Total	29	45	49	123

6 Length → p.137

1 ① 6m 17cm ② 6m 53cm
③ 9m 64cm ④ 9m 81cm

2 ① 4 ② 2,850 ③ 7000 ④ 3060

3 ① 1800m ② 3km 75m

4 ① 1km 150m ② 260m

5 ① 3km 990m ② 4km 20m
③ 1km 220m ④ 1km 450m

7 Circles and Spheres → p.138

1 Ⓐ center Ⓑ radius Ⓒ diameter

2 5cm

3 (omitted)

4 Ⓑ, Ⓒ, Ⓐ

5 15cm

6 14cm

8 Division with Remainders → p.139

1 ① 7 remainder 2 ② 6 remainder 6 ③ ○

2 ① 2 remainder 1, $4 \times 2 + 1 = 9$
② 2 remainder 2, $3 \times 2 + 2 = 8$
③ 9 remainder 1, $5 \times 9 + 1 = 46$
④ 4 remainder 5, $6 \times 4 + 5 = 29$
⑤ 4 remainder 4, $9 \times 4 + 4 = 40$
⑥ 8 remainder 3, $7 \times 8 + 3 = 59$

3 ① 7 children; 4 sheets remain
② 5 sheets for each child; 6 sheets remain

4 ① 7 ribbons; 1cm remains
② 6cm

5 8 plates

6 6 packs

[Let's deepen.]

Let's play at the amusement park! → p.140

① 10:05a.m.
② 54 minutes
③ 11:28 a.m.

Let's go shopping! → p.142

1 ① 670yen
② a melon (480yen) and an apple (130 yen)

2 ① 320yen ② 180yen
③ one 500-yen coin and two 10-yen coins

3 ① 370yen
② How to pay?…one 1000-yen bill, one 100-yen coin and three 10-yen coins
Money left… one 500-yen coin, three 100-yen coins, one 50-yen coin, one 10-yen coin

Is it a proper graph? → p.144

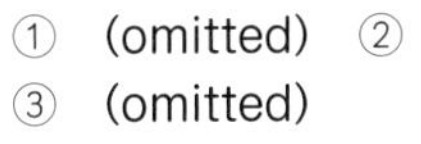
① (omitted) ②
③ (omitted)

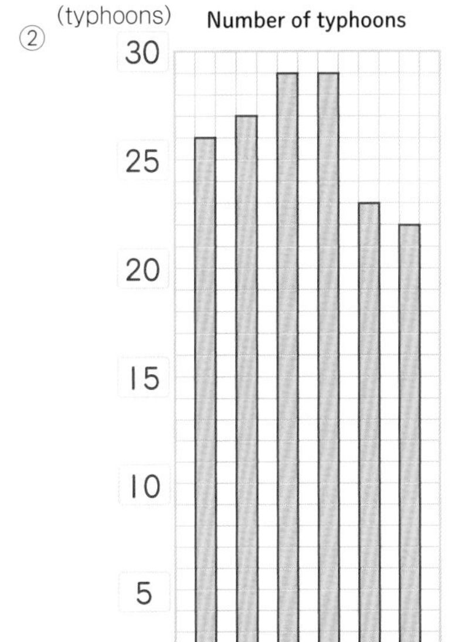

which we learned in this textbook

Multiplication Table

→ To be used in pp.13 ~ 22

Multiplier

Multiplicand

	0	1	2	3	4	5	6	7	8	9	10	11	12
0													
1		1	2	3	4	5	6	7	8	9			
2		2	4	6	8	10	12	14	16	18			
3		3	6	9	12	15	18	21	24	27			
4		4	8	12	16	20	24	28	32	36			
5		5	10	15	20	25	30	35	40	45			
6		6	12	18	24	30	36	42	48	54			
7													
8													
9													
10													
11													
12													

Let's make the rows of 10, 11, and 12.

Sara

Memo

Editors of Original Japanese Edition

[Head of Editors]

Shin Hitotsumatsu (Kyoto University), Yoshio Okada (Hiroshima University)

[Supervising Editors]

Toshiyuki Akai (Hyogo University), Toshikazu Ikeda (Yokohama National University), Shunji Kurosawa (The former Rikkyo University), Hiroshi Tanaka (The former Tsukuba Univ. Elementary School), Kosho Masaki (The former Kokugakuin Tochigi Junior College), Yasushi Yanase (Tamagawa University)

[Editors]

Shoji Aoyama (Tsukuba Univ. Elementary School), Kengo Ishihama (Showa Gakuin Elementary School), Hiroshi Imazaki (Hiroshima Bunkyo University), Atsumi Ueda (Hiroshima University), Tetsuro Uemura (Kagoshima University), Yoshihiro Echigo (Tokyo Gakugei Univ. Setagaya Elementary School), Hisao Oikawa (Yamato University), Hironori Osawa (Yamagata University Graduate School), Tomoyoshi Owada (Shizuoka University), Nobuhiro Ozaki (Seikei Elementary School), Masahiko Ozaki (Kansai Univ. Elementary School), Kentaro Ono (Musashino University), Hiroshi Kazama (Fukui University), Michihiro Kawasaki (Oita University), Miho Kawasaki (Shizuoka University), Yoshiko Kambe (Tokai University), Yukio Kinoshita (Kwansei Gakuin Elementary School), Tomoko Kimura (Tamon Elementary School, Setagaya City), Satoshi Kusaka (Naruto University of Education), Kensuke Kubota (Naruo Higashi Elementary School, Nishinomiya City), Itsushi Kuramitsu (The former University of the Ryukyus), Maiko Kochi (Kounan Elementary School, Toshima City), Chihiro Kozuki (Daiyon. Elementary School, Hino City), Goto Manabu (Hakuoh University), Michihiro Goto (Tokyo Gakugei Univ. Oizumi Elementary School), Hidenori Kobayashi (Hiroshima Univ. Shinonome Elementary School), Akira Saito (Shibata Gakuen University Graduate School), Masahiko Sakamoto (The former Tokoha University Graduate School), Junichi Sato (Kunitachigakuen Elementary School), Hisatsugu Shimizu (Keio Yochisha Elementary School), Ryo Shoda (Seikei University), Masaaki Sugihara (University of the Sacred Heart, Tokyo), Jun Suzuki (Gakushuin Primary School), Shigeki Takazawa (Shiga University), Chitoshi Takeo (Nanzan Primary School), Hidemi Tanaka (Tsukuba Univ. Elementary School), Toshiyuki Nakata (Tsukuba Univ. Elementary School), Hirokazu Nagashima (Daisan. Elementary School, Kokubunji City), Minako Nagata (Futaba Primary School), Kiyoto Nagama (Hiyagon Elementary School, Okinawa City), Satoshi Natsusaka (Tsukuba Univ. Elementary School), Izumi Nishitani (Gunma University), Kazuhiko Nunokawa (Joetsu University of Education), Shunichi Nomura (Waseda University), Mantaro Higuchi (Kori Nevers Gakuin Elementary School), Satoshi Hirakawa (Showa Gakuin Elementary School), Kenta Maeda (Keio Yokohama Elementary School), Hiroyuki Masukawa (University of the Sacred Heart, Tokyo), Shoichiro Machida (Saitama University), Keiko Matsui (Hasuike Elementary School, Harima Town), Katsunori Matsuoka (Naragakuen University), Yasunari Matsuoka (Matsushima Elementary School, Naha City), Yoichi Matsuzawa (Joetsu University of Education), Satoshi Matsumura (Fuji Women's University), Takatoshi Matsumura (Tokoha University), Kentaro Maruyama (Yokohama National Univ. Kamakura Elementary School), Kazuhiko Miyagi (Homei Elementary School Affiliated with J.W.U), Aki Murata (University of California, Berkeley), Takafumi Morimoto (Tsukuba Univ. Elementary School), Yoshihiko Moriya (The former Kunitachigakuen Elementary School), Junichi Yamamoto (Oi Elementary School, Oi Town), Yoshikazu Yamamoto (Showa Gakuin Elementary School), Shinya Wada (Kagoshima University), Keiko Watanabe (Shiga University)

[Reviser]

Katsuto Enomoto (The former Harara Elementary School, Kagoshima City)

[Reviser of Special Needs Education and Universal Design]

Yoshihiro Tanaka (Teikyo Heisei University)

[Cover]

Photo : Tatsuya Tanaka (MINIATURE LIFE)

Design : Ai Aso (ADDIX)

[Text]

Design : Ai Aso, Takayuki Ikebe, Hanako Morisako, Miho Kikuma, Sawako Takahashi, Katsuya Imamura, Satoko Okutsu, Kazumi Sakaguchi, Risa Sakemoto (ADDIX), Ayaka Ikebe

[Illustrations]

Lico, Kozue Gomita, Akira Sugiura, DOKOCHALUCHO, Kinue Naganawa, Mayumi Nojima, B-rise

[Photo・Observation data]

Showa Gakuin Elementary School, Ken Ogawa, DEPO LABO, Ritmo, Aflo, Pixta, Nissan Motor Co., Ltd., Itsukushima Shrine, Hiroshima Prefecture, Water Supply and Sewerage Sales Division,Fuji City, Tomihiro Art Museum, Kamikawagunhigashikawa cho,Hokkaido, Sewage and River Planning Division,Takatsuki City, Sewerage and River Bureau,Corporate Planning Division,Sapporo City, Waterworks Bureau,Corporate Planning Division,Kobe City, NASA